Additional Mathematics

Advanced FSMQ for OCR

By Vali Nasser

Copyright © 2014

E-book editions are also available for this title. For more information email:

valinasser@gmail.com

All rights reserved by the author. No part of this publication can be reproduced, stored in a retrieval system, or transmitted in any form or by any means, electronic, mechanical, photocopying, recording or otherwise, without the prior permission of the publisher and/or author.

New Edition: September 2014

ISBN-13: 978-1500984755

ISBN-10: 1500984752

> Every effort has been made by the author to ensure that the material in this book is up to date and in line with the requirements to pass the FSMQ Advanced exam by OCR at the time of publication. The author will also do his best to review, revise and update this material periodically as necessary. However, neither the author nor the publisher can accept responsibility for loss or damage resulting from the material in this book

INDICES AND SURDS 7

Fractional Indices 8

Rational and Irrational Numbers 8

Surds 10

ALGEBRA BASICS 11

Reminder: Multiplying positive and negative numbers. 11

Dividing positive and negative numbers. 11

Simplifying algebraic expressions 12

Multiplying out brackets. 13

SIMPLIFYING ALGEBRAIC FRACTIONS 16

Factorising 16

ALGEBRAIC SUBSTITUTION AND FORMULA 17

CHANGING THE SUBJECT 19

SOLVING LINEAR EQUATIONS 21

SOLVING WORD PROBLEMS USING ALGEBRA 24

BEARINGS 26

LINEAR EQUATIONS 27

WORKING OUT EQUATIONS OF 'NORMALS' AND 'PARALLEL' LINES 30

Finding parallel lines 31

SIMULTANEOUS EQUATIONS 32

SOLVING QUADRATIC EQUATIONS 36

SOLVING QUADRATIC INEQUALITIES 41

GRAPHS OF QUADRATIC EQUATIONS 43

CUBIC EQUATION 46

SOLVING EQUATIONS USING GRAPHICAL METHODS 47

Solving equations mathematically when one is linear and the other is quadratic: 47

EQUATION OF A CIRCLE 50

REMAINDER AND FACTOR THEOREM 52

Remainder Theorem 52

Factor Theorem 52

SOLVING CUBIC EQUATIONS 54

LINEAR PROGRAMMING 57

TRIG FOR RIGHT ANGLED TRIANGLES 60

TRIG FOR NON- RIGHT ANGLED TRIANGLES 63

GRAPHS OF TRIG FUNCTIONS	65
TRIG IDENTITIES	68
PYTHAGORAS' THEOREM	69
BINOMIAL EXPANSION	71
Binomial Theorem	72
Probability revisited	72
Binomial Probability Function	75
DIFFERENTIATION	78
INTEGRATION	81
Definite Integrals	82
Finding areas enclosed between curves	85
KINEMATICS	87
Differentiation and Integration in Kinematics	89
FORMULA SHEET	91
PRACTICE TEST 1 SECTION A	92
SECTION B	94
PRACTICE TEST 2 SECTION A	96
SECTION B	98

ANSWERS TO PRACTICE TEST 1   SECTION A AND B 100

ANSWERS TO PRACTICE TEST 2   SECTION A AND B 108

Introduction

This book will prepare you well for the Additional Mathematics exam set by OCR. Although it starts gently with a few reminders it builds up quickly to the level required in the various topics you can expect in the exam. The book assumes that you can achieve or have achieved a grade A or A* in GCSE maths.

The content covers the four essential areas that you will be tested on which include Algebra, Co-ordinate Geometry, Trigonometry and Calculus. It also includes some Kinematics to familiarize you with Applied Mathematics. This is a purely exam based test and has two sections. Section A typically has 10 questions totaling 52 marks and Section B has 4 questions totaling 48 marks. Naturally, you are expected to be familiar with using a scientific calculator which you are allowed to use in the exam.

The FSMQ Advanced (Free Standing Mathematics Qualification: Advanced) will prepare you well for doing mathematics at a higher level. This will be particularly true if you want to progress to AS and A2 with ease. It will also be useful for individuals who want to enhance their knowledge beyond GCSE.

About the Author

The author of this book has experience in both consultancy work and teaching. As a specialist mathematics teacher he has tutored and taught mathematics in schools as well as in adult education. The author's initial book 'Speed Mathematics Using the Vedic System' has a significant following and has been translated into Japanese and Chinese as well as German. His book 'Pass the QTS Numeracy Skills Test with Ease' has been very popular with teacher trainees and 'Pass the Numerical Reasoning Tests' is popular with graduates wanting to further their career. He hopes that his new book on Additional Mathematics will be helpful to pupils who need to revise the topics for the FSMQ Advanced exam.

Indices and Surds

You are probably already familiar with squares, square roots, cubes and cube roots. Powers, Indices/Index Numbers or exponents are simply the power by which a base number is raised. So just as 4^3 (4 cubed) means 4 to the power of 3, these 'powers' as mentioned earlier are also referred to as indices or index numbers. So 5^6 simply means 5 raised to the power of six. So in this case 5^6 means 5×5 5×5×5×5! (5 is called the base number and 6 is the power or index number) It is interesting to note that if you multiply two or more same base numbers with indices for example: $5^6 \times 5^3$ you simply add the indices to get 5^9 (5 to the power 9). Reason: 5^6 means 5×5×5×5×5×5 and 5^3 means 5×5×5 so $5^6 \times 5^3 = (5 \times 5 \times 5 \times 5 \times 5 \times 5) \times (5 \times 5 \times 5) = 5^9$

Similarly, for division, you simply subtract the indices. Consider $5^6 \div 5^3$. This means we need to work out $\frac{5X5X5X5X5X5}{5X5X5}$ which cancels down to 5×5×5 or 5^3 So you can see that when dividing the **same** base numbers with indices you simply subtract the indices.

The examples below will help you to consolidate the manipulation of the same base numbers with indices.

Example 1: $7^8 \times 7^4 \times 7^6 = 7^{18}$ (simply add the indices 8 + 4 + 6 = 18, hence the answer is: 7^{18})

Example 2: $9^{12} \div 9^5 = 9^7$ (simply subtract 5 from 12 to get 7, hence the answer is: 9^7)

Finally, you can also have negative indices which are inverses of the base numbers with the appropriate indices.

Example 1: $5^{-1} = \frac{1}{5}$ (Also called the reciprocal of 5).

Example 2: $6^{-2} = \frac{1}{6^2}$

Example 3: $5^{-6} = \frac{1}{5^6}$

Fractional Indices: Examples: (i) $2^{1/2}$ (2 to the power of $\frac{1}{2}$), (ii) $27^{1/3}$ (27 to the power of $\frac{1}{3}$. (It's worth noting that $2^{1/2}$ is the same as $\sqrt{2}$, $27^{1/3}$ is the same as $\sqrt[3]{3}$ and $8^{2/3}$ means (8 to the power of $\frac{2}{3}$.)

Rational and Irrational Numbers

Numbers can be either rational or irrational

Any number that can be written as p/q is a rational number, where p and q are whole numbers and q is not zero. Basically, the number is well defined and we know or can predict its pattern.

Examples of rational numbers are: $5 = \frac{5}{1}$, $-2 = \frac{-2}{1}$, $\frac{1}{2} = 0.5$, $\frac{2}{5} = 0.4$, $\frac{1}{3} = 0.33333$ (recurring)

$\frac{0}{5} = 0$, $\frac{4}{33} = 0.1212121212.....$

Examples of irrational numbers are: $\pi, \sqrt{2}, \sqrt{3}$ or $5\sqrt{7}$

For square roots and cube roots those with perfect roots are rational whereas others are irrational. So for example $\sqrt{25} = 5$ is rational, $\sqrt[3]{27} = 3$ is rational but as we saw earlier $\sqrt{2}$ is irrational.

For example π or $\sqrt{2}$ do not have a predictable pattern. We can approximate them but not calculate them exactly.

Summary for Indices:

Rules of indices:

(1) $a^m \times a^n = a^{m+n}$

(2) $a^m \div a^n = a^{m-n}$

(3) $(a^m)^n = a^{m \times n}$

(4) $a^0 = 1$

(5) $a^{-1} = \dfrac{1}{a}$

(6) $a^{-m} = \dfrac{1}{a^m}$

Surds

Surds are simply expressions with irrational square roots. There are some useful rules associated with them.

1: $\sqrt{2} \times \sqrt{2} = \sqrt{4} = 2$

2: $\sqrt{3} \times \sqrt{2} = \sqrt{6}$

3: $\dfrac{\sqrt{6}}{\sqrt{2}} = \sqrt{\dfrac{6}{2}} = \sqrt{3}$

4: $(\sqrt{p} + \sqrt{q})^2 = (\sqrt{p} + \sqrt{q}) \times (\sqrt{p} + \sqrt{q}) = p + 2\sqrt{pq} + q$

5. $(p + \sqrt{q})(p - \sqrt{q}) = p^2 + p\sqrt{q} - p\sqrt{q} - q = p^2 - q$

6. $\dfrac{2}{\sqrt{3}}$ (Multiply top and bottom by $\sqrt{3}$) so we have: $\dfrac{2}{\sqrt{3}} \times \dfrac{\sqrt{3}}{\sqrt{3}} = \dfrac{2\sqrt{3}}{3}$

   (we call this rationalising the denominator)

7. $\dfrac{2}{1-\sqrt{3}}$ to simplify this we need to <u>rationalise</u> the denominator.

   To do this we simply multiply top and bottom by $(1 + \sqrt{3})$

   So we have, $\dfrac{2}{1-\sqrt{3}} = \dfrac{2}{1-\sqrt{3}} \times \dfrac{1+\sqrt{3}}{1+\sqrt{3}} = \dfrac{2(1+\sqrt{3})}{1-3} = \dfrac{2(1+\sqrt{3})}{-2} = -(1 + \sqrt{3})$

Note: <u>Leaving your answers as surds is quite respectable, since you can't work out the exact answer on a calculator!</u>

Algebra basics

$x(x+y) = x^2 + xy$

$x^2(x + x^2 + y) = x^3 + x^4 + x^2 y$

In general, a × a × a× a ....(n times) $= a^n$

You also need to know these algebraic rules for the multiplication and division of positive and negative numbers.

Reminder: Multiplying positive and negative numbers.

(+) × (+) = +   (a plus number times a plus number gives us a plus number)

(+) × (−) = −   (a plus number times a minus number gives us a minus number)

(−) × (+) = −   (a minus number times a plus number gives us a minus number)

(−) × (−) = +   (a minus number times a minus number gives us a plus number)

Dividing positive and negative numbers.

(+) ÷ (+) = + (a plus number divided by a plus number gives us a plus number)

(+) ÷ (−) = − (a plus number divided by a minus number gives us a minus number)

(−) ÷ (+) = − (a minus number divided by a plus number gives us a minus number)

(−) ÷ (−) = + (a minus number divided by a minus number gives us a plus number)

Summary: <u>For both multiplication and division, like signs give us a plus sign and unlike signs give a minus sign</u>

Also when adding and subtracting it is worth knowing that **when you add two minus numbers you get a bigger minus number.**

Example 1: $-4 - 6 = -10$

When you add a plus number and a minus number you get the sign corresponding to the bigger number as shown below:

Example 2: $+6 - 9 = -3$, whereas, $-6+9 = 3$

When you subtract a minus from a plus or minus number you need to note the results as shown below:

Example 3: $6 -(-3)$ we get $6+3 = 9$ (since $-(-3) = +3$)

Example 4: $7 -(+3)$ we get $7 - 3 = 4$ (since $-(+3) = -3$)

In this case note that $-(-) = +$. Also, $+(-) = -$ and $-(+) = -$.

Simplifying algebraic expressions

Example 1: Simplify $3m^2 + 4y^3 + 4m^2 - 5y^3$

Method: We add and subtract like terms.

Now $3m^2 + 4m^2 = 7m^2$ and $4y^3 - 5y^3 = -y^3$

Hence, $3m^2 + 4y^3 + 4m^2 - 5y^3 = 7m^2 - y^3$

Example 3: Simplify $ax^2 \times a^4 x^3$

Method: Multiply the two expressions and add the indices for similar terms:

This means $ax^2 \times a^4 x^3 = a^5 x^5$

Example 4: Simplify $\dfrac{y^3}{x^2} \div \dfrac{y^2}{x}$

Method: Using the rules of indices we get : $\dfrac{y^3}{x^2} \div \dfrac{y^2}{x} = \dfrac{y^3}{x^2} \times \dfrac{x}{y^2} = \dfrac{y}{x}$

(**N.B**. In the above example you can of course cancel down to get the same answer)

Multiplying out brackets.

Example 1: Expand and simplify $3(2x+5) + 4(2x+7)$

Method: Multiply 3 by each term in the first bracket then 4 by each term in the second bracket. The final step is to simplify by collecting up the like terms.

$3(2x+5) + 4(2x+7) = 6x + 15 + 8x + 28 = 14x + 43$

Example 2: Work out $(2x+3)(2x+4)$

When we have to multiply out two brackets we have to multiply each term in the first bracket by each term in the second bracket. We then simplify the resulting expression as before. An easy way to multiply out two brackets is to use the grid method as shown below:

First put each of the terms of each bracket on the outside grid as shown

×	2x	+3
2x		
+4		

Step2: Multiply each outside term together. So that for example $2x \times 2x = 4x^2$. The other results are shown inside the grid.

×	2x	+ 3
2x	$4x^2$	+ 6x
+ 4	8x	+12

After multiplying out the terms, the answer is found by adding all the terms inside the grid and simplifying the resulting expression.

So we have, $4x^2 + 6x + 8x + 12$ (These are all the terms inside the grid)

Finally, $4x^2 + 6x + 8x + 12 = 4x^2 + 14x + 12$

Another example will help consolidate the process:

Multiply out $(2x - 3)(3x + 2)$

Put the terms of each bracket on the outside of the grid as shown

×	2x	−3
3x	$6x^2$	−9x
+2	4x	−6

Collecting up all the terms inside the grid we have:

$6x^2 - 9x + 4x - 6$

Now simplify, which gives us $6x^2 - 5x - 6$

Another way of expanding brackets

Example 1: Expand $(x + 3)(x + 2)$

(Multiply the first term of the first bracket by the second bracket and then multiply the second term of the first bracket by the second bracket. Finally simplify the expression.)

So $(x + 3)(x + 2) = x(x + 2) + 3(x + 2) = x^2 + 2x + 3x + 6 = x^2 + 5x + 6$

Example 2: Expand $(2x - 1)(x - 2)$

This equals $2x(x - 2) - 1(x - 2) = 2x^2 - 4x - x + 2 = 2x^2 - 5x + 2$

Example 3: Expand $(y^2 + x)(y - x^2)$

$= y^2(y - x^2) + x(y - x^2) = y^3 - y^2x^2 + xy - x^3$

Harder example:

Expand and simplify the expression $(x^2 + 2x + 1)(2x^2 + 3x - 2)$

The principle is the same. Take each term in the first bracket and multiply it out by the second bracket. Finally simplify as much as you can.

So $(x^2 + 2x + 1)(2x^2 + 3x - 2) = x^2(2x^2 + 3x - 2) + 2x(2x^2 + 3x - 2) + 1(2x^2 + 3x - 2) = 2x^4 + 3x^3 - 2x^2 + 4x^3 + 6x^2 - 4x + 2x^2 + 3x - 2$

This simplifies to: $2x^4 + 7x^3 + 6x^2 - x - 2$

Difference of two squares

Something useful to remember is the <u>difference of two squares</u>:

$$p^2 - q^2 = (p + q)(p - q)$$

Since $(p + q)(p - q) = p(p - q) + q(p - q) = p^2 + pq - pq - q^2 = = p^2 - q^2$

So for example: $16x^2 - 9y^2 = (4x - 3y)(4x + 3y)$

Simplifying Algebraic Fractions

Example 1: Simplify $\frac{1}{x+3} + \frac{2}{3}$

Method: First find the common denominator which is $3(x + 3)$

Then treat it like you were simplifying a fraction

$$\frac{1}{x+3} + \frac{2}{3} = \frac{1\times 3 + 2(x+3)}{3(x+3)} = \frac{3+2x+6}{3(x+3)} = \frac{9+2x}{3(x+3)}$$

Example 2: Simplify $\frac{2}{x-3} - \frac{1}{5}$

As before $\frac{2}{x-3} - \frac{1}{5} = \frac{5\times 2 - 1(x-3)}{5(x-3)} = \frac{10-x+3}{5(x-3)} = \frac{13-x}{5(x-3)}$

Example 3: Simplify $\frac{2}{x-2} + \frac{3}{x+2}$

Using the method before we get:

$$\frac{2}{x-2} + \frac{3}{x+2} = \frac{2(x+2)+3(x-2)}{(x-2)(x+2)} = \frac{2x+4+3x-6}{x^2-4} = \frac{5x-2}{x^2-4}$$

Note: $(x+2)(x-2) = x^2 + 2x - 2x - 4 = x^2 - 4$

Factorising

Example 1: Factorise: $3x^2 - 6xy$

$= 3x(x - 2y)$ (find the common factor which is $3x$ in this case)

Example 2: Factorise: $3t^2y - 9t^3$

$= 3t^2(y - 3t)$

Algebraic Substitution and Formula

This is the process of substituting numbers for letters and working out value of the corresponding expression. Some examples that will clarify the process.

Example 1: If k=6 and t=8 work out 2(4k–2t) +kt

Substituting the values of k and t we have:

2(4 × 6–2 × 8) + 6 ×8

=2 × (24 – 16) +48 = 2 ×8 +48 =16+48 =64

So 2(4k – 2t) + kt = 64

Example 2: If t=9 and u= 6 work out $3t^2$ -5u

Substituting appropriately we get:

3 ×9^2 - 5 ×6 = 3 × 81–30 =243 – 30 =213

(Notice, we use the BIDMAS rule to work out the square first and then do the multiplication)

So, $3t^2$ -5u =213

Formula

A formula describes the relationship between two or more variables. You have seen some examples above already. Now let us consider some practical examples.

Example:

The formula for converting the temperature from Celsius to Fahrenheit is given by the formula: F= $\frac{9}{5}$ C +32   (where C is the temperature in degrees Centigrade)

If the temperature is 10 degrees Celsius then what is the equivalent temperature in Fahrenheit?

Using the formula $F = \frac{9}{5}C + 32$, and substituting 10 in place of C, we have $F = \frac{9}{5} \times 10 + 32 = \frac{90}{5} + 32 = 18 + 32 = 50$. Hence, 10 degrees centigrade = 50 degrees Fahrenheit

Explanation of working out above: Remember we multiply and divide before adding and subtracting) There are no brackets to worry about. When working out $\frac{9}{5} \times 10 + 32$, multiply 9 by 10 to get 90, divide this by 5 to get 18, finally add 18 and 32 together to get 50

Example 3: Convert 68 degrees Fahrenheit to degrees Celsius. The formula for converting the temperature from Fahrenheit to Celsius is given by:

$C = \frac{5}{9}(F-32)$, So to change 68 degrees Fahrenheit to degrees Celsius we can substitute for F in the formula $C = \frac{5}{9}(F-32)$, $C = \frac{5}{9}(68-32) =$

$\frac{5}{9} \times 36 = 5 \times 4 = 20$. Hence, 68 degrees Fahrenheit = 20 degrees Celsius

Explanation of the working out above: Using BIDMAS we work out the bracket first. This gives us 68-32 = 36. We now divide this by 9 and multiply by 5. Clearly 36÷9 = 4 and finally 5×4 = 20

Changing the subject

Changing the subject of a formula (re-arranging formulas)

Example 1: In the formula $a = bx + c$ make x the subject

Method: Apply the same rules as you would to equations

In this case subtract c from both sides to get $a - c = bx$

Now divide both sides by b to get $\dfrac{a-c}{b} = x$

In other words $x = \dfrac{a-c}{b}$

Example 2: In the formula $\dfrac{ay^2}{b} + m = k$, make y the subject

Method:

Step 1: Subtract m from both sides to get $\dfrac{ay^2}{b} = k - m$

Step 2: Multiply both sides by 'b' to get $\dfrac{ay^2}{\cancel{b}} \times \cancel{b} = b(k - m)$

(The 'b' s on the left hand side cancel)

Hence we now have: $ay^2 = b(k - m)$

Step 3: divide both sides by 'a' (to cancel the 'a' on the left hand side)

We now have $y^2 = \dfrac{b(k-m)}{a}$

Step 4: Take the square root of both sides to get $y = \sqrt{\dfrac{b(k-m)}{a}}$

So we finally find that $y = \sqrt{\dfrac{b(k-m)}{a}}$

Example 3: In the formula $\dfrac{a}{1+t^2} = b + c$, make t the subject

Step 1: Multiply both sides by $1 + t^2$ to get $a = (b + c)(1 + t^2)$

Step 2: Divide both sides by $(b + c)$ to get: $\dfrac{a}{b+c} = 1 + t^2$

Step 3: Subtract '1' from both sides to get: $\dfrac{a}{b+c} - 1 = t^2$

Step 4: This simplifies to $\dfrac{a-1(b+c)}{b+c} = t^2$, which simplifies to $\dfrac{a-b-c}{b+c} = t^2$

Step 5: Take the square root of both sides: $\sqrt{\dfrac{a-b-c}{b+c}} = t$

So finally we have $t = \sqrt{\dfrac{a-b-c}{b+c}}$

Solving Linear Equations

Example 1: Solve the equation 5x − 1 = 2x +8

First add 1 to both sides, which gives:

5x = 2x +9

Now subtract 2x from both sides to give 3x = 9

Finally divide both sides by 3 to get x=3.

(Notice each step simplifies the equation further)

Example 2: Solve the equation 5(2x +1) =4(2x +1)

To solve this first multiply out the bracket which gives:

10x +5 = 8x +4

(Multiply each term outside the bracket by each term inside the bracket)

Now subtract 5 from both sides, which gives:

10x =8x −1

Now subtract 8x from both sides, which gives:

2x = −1

Finally, divide both sides by 2 to get x= −1/2 or −0.5

Example 3: Solve the equation $\frac{2x}{3} + 5 = 7$

We can simplify this to $\frac{2x}{3} = 2$ (by subtracting 5 from both sides)

Now multiply both sides by 3 to get the expression below:

$2x = 6$, so $x = 3$

Example 4

Solve the equation $\sqrt{4 - \frac{x+3}{3x+2}} = 3$

Although this might look complicated the basic rule is whatever you do to one side you must do the same to the other.

Step 1: Square both sides so we get $4 - \frac{x+3}{3x+2} = 9$

Step 2: Cross –multiply everything by the denominator $(3x + 2)$

We get: $4(3x + 2) - (x + 3) = 9(3x + 2)$

Simplify to get $12x + 8 - x - 3 = 27x + 18$

Simplify further to get $11x + 5 = 27x + 18$

Subtract 18 from both sides to get $11x - 13 = 27x$

Now subtract $11x$ from both sides to get $-13 = 16x$

Which is the same as $16x = -13$, this means $x = \frac{-13}{16}$

Solving linear equations with inequalities

Example 1: Solve the inequality $2x + 5 > 9$

This simply says $2x + 5$ is greater than 9. To find x still use the rules of a simple equation. That is, whatever you do to one side you must do to the other.

If $2x + 5 > 9$, then $2x > 4$ (by taking away 5 from both sides)

Now, divide both sides by 2 to get $x > 2$. Our answer for x is all values greater than 2.

Example 2: Solve the inequality $2(5x - 1) \geq 3x + 14$

Method:

This simplifies to $10x - 2 \geq 3x + 14$

Subtract 3x from both sides to get $7x - 2 \geq 14$

Add 2 to both sides to get $7x \geq 16$

Dividing both sides by 7 to get $x \geq \frac{16}{7} \implies x \geq 2\frac{2}{7}$

Example 3: Solve the inequality $4 - 2x < 16$

$\implies -2x < 12 \implies -x < 6 \implies x > -6$

(Note: <u>In inequalities when you divide both sides by -1 you also change the sign of the inequality</u>)

Example 3: Show the inequality $-2 < x \leq 2$ by way of a number line.

The answer is shown below:

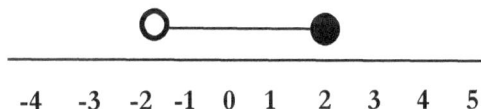

(Note the convention that a dark shaded circle implies 2 is included because $x \leq 2$ and at -2 unshaded (open) circle implies -2 is not included because it is <)

Solving Word Problems using Algebra

Examples:

(1) Fatima and Louise have £350 between them. Louise has £80 less than Fatima. How much do they each have?

Method: Let the amount Fatima has be represented by x
Hence, Louise has x – 80. We know that the sum of the two amounts = £350. That is x + x – 80 =350. Simplifying, we get 2x – 80 =350. Now add 80 to both sides so we have 2x - 80 +80 = 350 + 80. Which means 2x = 430, or x = 215. This means Fatima has £215 and Louise has £135 (Since Louise has £80 less than Fatima)

(2) The cost of a coat after a 20% discount is £85. What was its original price?

Method: Let the original price be £x. This means x – 20% of x = 85. Or x – 0.2x = 85, which simplifies to 0.8x =85. Now divide both sides by 0.8. So we get x = 85÷0.8 = 106.25. Hence the original price is £106.25

(3) The area of a rectangle is $162m^2$. The length of the rectangle is two times the width. What is the length and width of the rectangle?

Method: Let the width =w, hence the length = 2w. We know that the area of a rectangle is length × width = 2w×w = $2w^2$. The area of the rectangle is given as $162m^2$. Hence, $2w^2$ = 162. Dividing both sides by 2, we get w^2= 81. Hence w = $\sqrt{81}$ =9. So the width is 9m and the length is 18m.

(4) John's annual salary is $\frac{3}{4}$ of Hilary's salary. Hilary's salary is twice Betty's. The total salary between them is $450,000. How much did each of them earn?

Method: Let Hilary's salary be x (in dollars). Hence, John's salary is $\frac{3}{4}$x. Also, since Hilary earns twice as much as Betty, then Betty earns half of

Hillary's $= \frac{1}{2}x$. Finally, we know that $x + \frac{3}{4}x + \frac{1}{2}x = \$450,000$, simplifying $2\frac{1}{4}x = 450,000$. Or, $\frac{9}{4}x = 450,000$. This means $9x = 1,800,000$ or $x = 200,000$. So Hilary earns $200,000, John earns $150,000 (three quarters of Hilary's amount) and Betty earns $100,000 (half of Hilary's salary)

(5) The sum of two numbers is 30 and the difference between them is 8. What are the two numbers?

Method: Let the unknown numbers be x and y. This means $x + y = 30$, and $x - y = 8$, If we add the above two equations we get $2x = 38$ (The y's cancel). Hence $x = 19$ and $y = 11$

(6) The second number is $\frac{3}{4}$ of the first number. The sum of two numbers is 5.25. What are the two numbers?

Method: The sum of the two numbers is 5.25. Let one of the numbers be x. Hence, $x + \frac{3}{4}x = 5.25$. Simplifying, we get $1.75x = 5.25$, dividing both sides by 1.75 we get $x = 3$. So $\frac{3}{4}x = 0.75 \times 3 = 2.25$. So the two numbers are 3 and 2.25.

Bearings

Bearings are measured clockwise from the **North line** to the line joining the two points as shown below:

Example: The bearing of A from B is measured clockwise from the North line and in this case is 75°

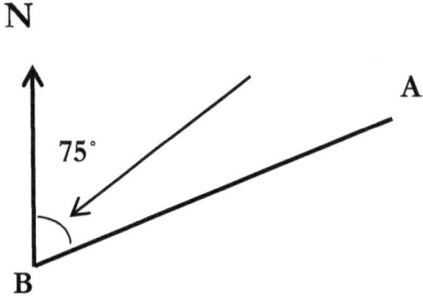

Example: Find the bearing of Q from P.

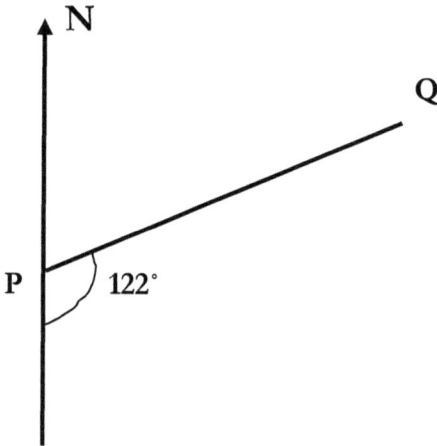

Method: Since you are going from P to Q we need to find the angle between the North line and Q.

This is 180° - 122° = **58°**

Linear equations

These are of the form y = mx + c

where m is the gradient or slope and c is the value of y when x = 0

Example 1: y = 3x - 1

The graph is shown below.

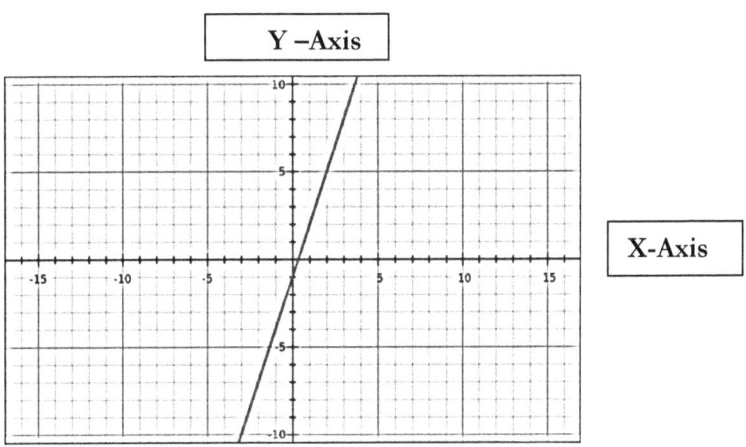

You can see that the equation y = 3x – 1 crosses the y –axis at y = - 1 (this is the called the intercept)

In other words when x = 0, y = 3×0 – 1 = -1

The '3' in the 3x bit refers to the gradient or the slope of the graph.

So in general a linear equation is of the form y = mx + c, where m is the gradient and c is the value of y when x = 0

Example 1

Plot the equation y = 2x - 3 for values of x = -2 to +2 by completing the table below first. The plotted graph is shown below the completed table.

x	-2	-1	0	1	2
2x – 3	-7	-5	-3	-1	1
y	-7	-5	-3	-1	1

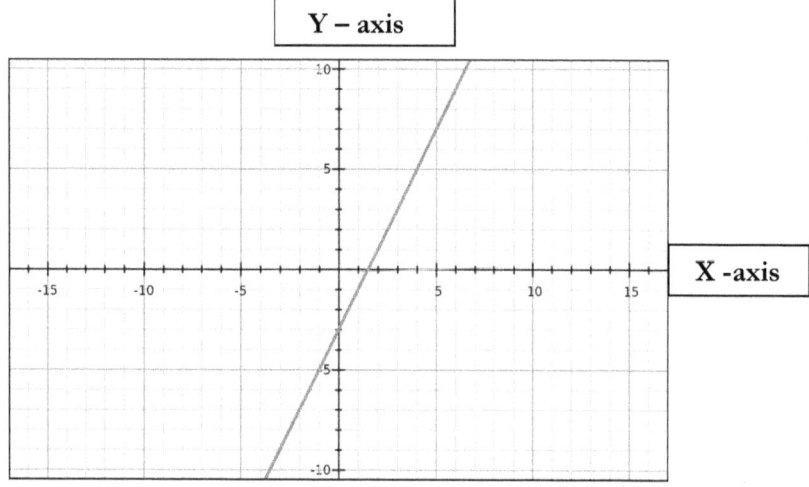

Example 2

The co-ordinates P (2, 3) and Q (4, 6) lie on a straight line.

(i)   Find the mid- point R of the co-ordinates P & Q
(ii) Find the gradient of the straight line
(iii) Find the equation of the straight line

(i) The mid-points are simply the average of the x co-ordinates and the y co-ordinates.

Mid –point for the x co-ordinate = $\frac{x1+x2}{2} = \frac{2+4}{2} = 3$

Similarly, the mid-point of the y co-ordinate = $\frac{y1+y2}{2} = \frac{3+6}{2} =$ 4.5

Hence R, the mid-point of PQ is (3, 4.5)

(ii) The gradient of two points that lie on a straight line = $\frac{Difference\ in\ y\ co-ordinates}{Difference\ in\ x\ co-ordinates} = \frac{6-3}{4-2} = \frac{3}{2} = 1.5$

(iii) The equation of a straight line is given by y =mx + c

Since m = 1.5, then the equation is y = 1.5x + c
We also know that it goes through P, Q and R. So we can choose any of these to find the value of C. Let us choose P(2, 3) so the equation is now:
3 = 1.5×2 + C, Hence 3 = 3 + C so in this case C =0
So the equation is y = 1.5x

(iv) You can also find the equation of a straight line if you know the co-ordinates of a point it goes through and its gradient by using the formula: y – y1 = m(x – x1). (where x1, y1 are the co-ordinates of the point the line goes through and m is the gradient).

Working out equations of 'Normals' and 'Parallel' lines

If two lines are perpendicular to each other, then their gradients, m1 & m2 when multiplied together = – 1. That is $m_1 m_2 = -1$ (The line that is perpendicular to given line is called a 'normal')

Example 1

Given the equation $y = 3x - 1$. Find the equation of a line that is perpendicular to it and goes through the co-ordinates P(1, -2)

Method: Gradient of the line $y = 3x - 1$ is 3 (since if $y = mx + c$, then m is the gradient). Clearly if $m1 \times m2 = -1$ then $3 \times m2 = -1$. This means $m2 = -\frac{1}{3}$.

Finally if this line goes through P (1, -2) then its equation can be found by using the fact that $y - y1 = m(x - x1)$ where x1, y1 are the co-ordinates it goes through. This means the equation of the **normal** or the **line that is perpendicular to** $y = 3x - 1$ is given by $y - (-2) = -\frac{1}{3}(x - 1)$. (We simply substituted the co-ordinates of P(1,-2) in the equation $y = 3x - 1$)

$\Rightarrow y + 2 = -\frac{1}{3}(x - 1) \Rightarrow 3y + 6 = -x + 1 \Rightarrow 3y + x = -5$

$\Rightarrow y = -\frac{x}{3} - \frac{5}{3}$ or $y = -\frac{x}{3} - 1\frac{2}{3}$

Alternative method for finding the perpendicular line:

Since the equation of a straight line is given by y = mx + c, then the equation of the perpendicular line is $y = -\frac{1}{3}x + c$

We can find c by substituting P(1, -2) in this equation. Substituting for x and y we get: $-2 = -\frac{1}{3} \times 1 + c$. This means $c = -2 + \frac{1}{3} = -1\frac{2}{3}$; hence $y = -\frac{1}{3}x - 1\frac{2}{3}$

Finding parallel lines

This is more straightforward **since the gradient of parallel lines are the same.**

Example: Find the line that is parallel to $y = \frac{1}{2}x + 3$ which passes through the point (2, -1)

Method: We know that the equation of a line is given by y = mx + c

Since the gradient of this parallel line is the same, the equation of this parallel line is: $2 = \frac{1}{2} \times (-1) + c \implies c = 2 + \frac{1}{2} = 2\frac{1}{2}$. So the equation of the **parallel line** which goes through (2, -1) is $y = \frac{1}{2}x + 2\frac{1}{2}$

Simultaneous Equations

We saw earlier that simple equations allow us to solve problems involving one unknown. When you have to solve problems involving more than one unknown you need more than one equation to solve these.

A simultaneous equation with two variables (meaning two unknowns), say x and y, typically involves two equations. The problem is then to find the values of x and y which satisfies the equations at the same time. Another way of saying simultaneous is 'at the same time'.

We will first consider one traditional method of solving simultaneous equations.

Example 1: Solve the simultaneous pair: $2x + 4y = 5$

$$3x + 5y = 9$$

First let us understand the problem. The problem is to find the values of x and y such that the equations are true.

Method 1:

Consider the substitution method:

We will express x in term of y in the first equation and substitute for x in the second equation. We have:

$2x + 4y = 5$

$3x + 5y = 9$

From the first equation we have $2x = 5 - 4y$ (subtract 4y from both sides)

So $x = 2.5 - 2y$ (we get this by dividing both sides of the previous expression by 2.)

Now substitute $x = 2.5 - 2y$ in the second equation, which gives us:

$3(2.5 - 2y) + 5y = 9$

So, $7.5 - 6y + 5y = 9$

So, $7.5 - y = 9$

So, $7.5 - 9 = y$

Which gives $y = -1.5$

Now we need to find x. We can substitute the value of y in equation 1 to find x.

The first equation is $2x + 4y = 5$

Substituting for y, we get $2x + 4 \times (-1.5) = 5$

which means $2x - 6 = 5$, So $2x = 11$, hence $x = 5.5$

Check

We can check in equation 1 to see if the values we found satisfy the equation.

The first equation is: $2x + 4y = 5$

Substituting, $x = 5.5$ and $y = -1.5$ we get:

$2 \times 5.5 + 4 \times (-1.5) = 11 - 6 = 5$ as required.

Method 2: Eliminate one of the variables

Consider the simultaneous pair of equations as before:

$2x + 4y = 5$ (1)

$3x + 5y = 9$ (2)

Try and make the x or the y terms the same and then add or subtract the equations to eliminate one of the variables

Suppose we make the 'x' term the same. We multiply equation (1) by 3 and equation (2) by 2. The new equations are now shown below:

$6x + 12y = 15$ (3)

$6x + 10y = 18$ (4)

Now subtract (4) from (3) and we get:

6x – 6x + 12y – 10y = 15 – 18

2y = -3 hence y = -3/2 =-1.5

Now substitute y = -1.5 in equation (1) and we get:

2x + 4×(-1.5) = 5 which means 2x – 6 = 5 or 2x = 11 so x = 11/2 or 5.5

Hence as before x =5.5 and y = -1.5

There is another method which involves drawing graphs of the two equations and finding the point at which they intersect.

Simultaneous equations with three unknowns:

Solve the simultaneous equations:

2x +3y + z = 4 ……………..(1)

3x - 3y + 2z = 2 ……………. (2)

2x + 3y - z = 2 …………….(3)

If we add (1) & (2) and adding (2) & (3) we can eliminate y.

So we now have 5x + 3z = 6 ………..(4)

$$5x + z = 4 \ldots\ldots\ldots(5)$$

We can now subtract (5) from (4) to get 2z = 2 $\implies$ z = 1

Substitute z =1 in (5) $\implies$ 5x + 1 = 4 $\implies$ x = $\frac{3}{5}$

Finally substitute z =1 and x = $\frac{3}{5}$ in (1) to get $\frac{6}{5}$ + 3y + 1 =4

$\implies$ 3y = 3 - $\frac{6}{5}$ $\implies$ 3y = $\frac{15-6}{5}$ $\implies$ 3y = $\frac{9}{5}$ $\implies$ y = $\frac{3}{5}$

Hence the solution to the simultaneous equations with three unknowns in this case are $x = \dfrac{3}{5}$, $y = \dfrac{3}{5}$ and $z = 1$

Solving Quadratic Equations

For the general quadratic equation $ax^2 + bx + c = 0$

The formula for solving the equation is given by: $x = \dfrac{-b \pm \sqrt{b^2 - 4ac}}{2a}$

We can show that this is true by the method of completing the square as shown below.

Consider the general quadratic equation $ax^2 + bx + c = 0$

Dividing through by 'a' we get:

$$x^2 + \dfrac{b}{a}x + \dfrac{c}{a} = 0$$

Now we use the method of completing the square

First halve the middle term coefficient and then square the expression on the left hand side as shown below:

$$(x + \dfrac{b}{2a})^2 = x^2 + \dfrac{b}{a}x + \dfrac{b^2}{4a^2}$$

Adjusting to get the original expression, we have:

$$(x + \dfrac{b}{2a})^2 - \dfrac{b^2}{4a^2} + \dfrac{c}{a} = x^2 + \dfrac{b}{a}x + \dfrac{c}{a}$$

We can write $(x + \dfrac{b}{2a})^2 - \dfrac{b^2}{4a^2} + \dfrac{c}{a} = 0$

$$(x + \dfrac{b}{2a})^2 = \dfrac{b^2}{4a^2} - \dfrac{c}{a}$$

Simplifying the right hand side we get:

$$(x + \frac{b}{2a})^2 = \frac{b^2}{4a^2} - \frac{4ac}{4a^2}$$

$$(x + \frac{b}{2a})^2 = \frac{b^2 - 4ac}{4a^2}$$

$$x + \frac{b}{2a} = \pm \frac{\sqrt{b^2 - 4ac}}{2a}$$

$$x = -\frac{b}{2a} \pm \frac{\sqrt{b^2 - 4ac}}{2a}$$

$$x = \frac{-b \pm \sqrt{b^2 - 4ac}}{2a}$$

Example:

Solve the equation $2x^2 - 5x + 2 = 0$ using the quadratic formula.

Method:

When the above equation is compared to the general equation

$ax^2 + bx + c = 0$

We can see that a = 2, b= -5 and c =2

Since, $x = \frac{-b \pm \sqrt{b^2 - 4ac}}{2a}$

By substituting the above values we can see that:

$$x = \frac{-(-5) \pm \sqrt{(-5)^2 - 4 \times 2 \times 2}}{2 \times 2}$$

$$x = \frac{5 \pm \sqrt{25-16}}{4}$$

$$x = \frac{5 \pm \sqrt{9}}{4}$$

$$x = \frac{5 \pm 3}{4} = \frac{8}{4} \text{ or } \frac{2}{4}$$

Hence $x = 2$ or $\frac{1}{2}$

(Note: The formula method is particularly useful if you find it hard to factorise or if a quadratic expression cannot be factorised)

Solving quadratic equations using factorisation when possible:

Example 1: Solve the equation $x^2 + 5x + 6 = 0$

We can factorise the above quadratic equation as $(x + 3)(x + 2) = 0$

This means either $x + 3 = 0 \implies x = -3$ or $x + 2 = 0 \implies x = -2$

Example 2: Solve the quadratic equation $2x^2 - 5x + 2 = 0$

We can write the above equation as $(2x - 1)(x - 2) = 0$

(You can do this by trial and error with a little bit of common sense)

For example the only way to get $2x^2$ is by having x and 2x in the two brackets. Also the only way to get + 2 as the last term is to have +1 and +2 or -1 and -2. Finally, as the middle term is -5x the factors have to be $(2x - 1)(x - 2)$

So if $(2x - 1)(x - 2) = 0$ this means either $2x - 1 = 0$ or $x - 2 = 0$

If $2x - 1 = 0 \implies 2x = 1$ and $x = \frac{1}{2}$ and if $x - 2 = 0 \implies x = 2$

Hence the solution to the quadratic equation $2x^2 - 5x + 2 = 0$ is

Either $x = \frac{1}{2}$ or $x = 2$

Things to note in quadratic equations and the quadratic formula:

(1) $ax^2 + bx + c = 0$ is a quadratic equation providing 'a' is not zero.
(2) There are two solutions (or roots) to a quadratic equation
(3) The roots are real so long as in the formula $x = \dfrac{-b \pm \sqrt{b^2 - 4ac}}{2a}$
the bit inside the square root is > 0. The bit inside the square root, that is $b^2 - 4ac$, is called the <u>discriminant</u>. Note if $b^2 - 4ac = 0$, there is only one real root
(4) When $b^2 - 4ac < 0$, then the roots are not real.

Below are examples of equations with two solutions (two roots), one solution (one root) and no solution (no real roots)

(a) The equation $x^2 + 2x - 15 = 0$, has two real roots, x= -5 and x = 3 as shown in the graph below

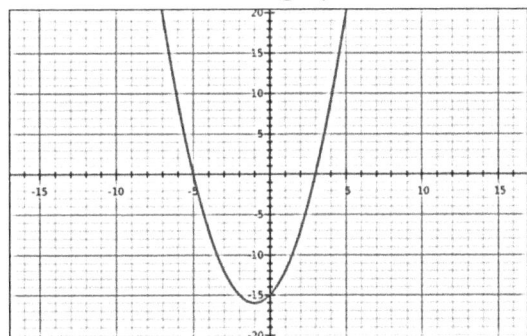

(b) The equation $x^2 - 6x + 9 = 0$ has one real root at x = 3 as shown below

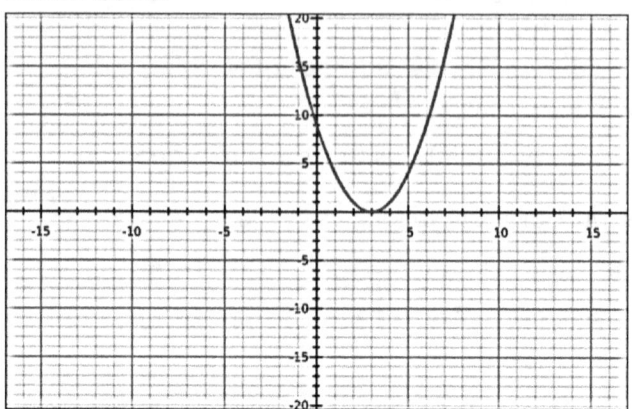

(c) The equation $x^2 - 6x + 12 = 0$ has no real roots as the parabola does not intersect the x-axis at any point.

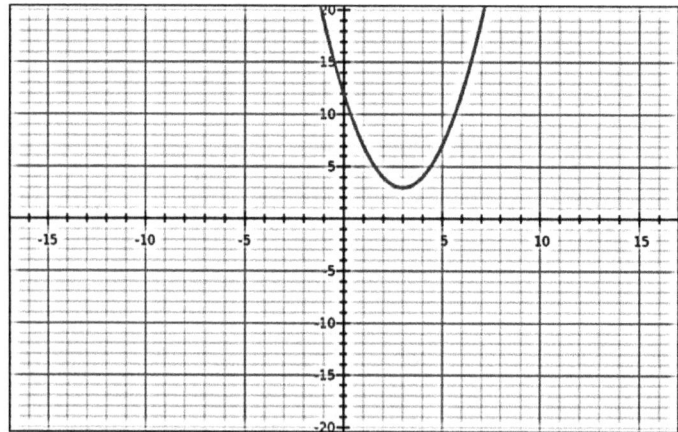

Solving Quadratic Inequalities

Example 1: Solve the quadratic inequality $-x^2 + x + 8 \geq 2$

First re-write this as $-x^2 + x + 6 \geq 0$

Plot its graph and find the values that satisfy this inequality. Namely, values of x, when $y \geq 0$.

First let us see if we can simplify and factorise the equation $-x^2 + x + 6 = 0$. Multiplying through by -1 we get $x^2 - x - 6 = 0. \Rightarrow x^2 - x - 6 = 0$

$\Rightarrow$ $(x - 3)(x + 2) = 0 \Rightarrow$ **x = 3 or -2**. (Also note that in the initial equation the coefficient of x^2 was negative this implies that the graph is inverted 'U' shaped as shown below)

y- axis

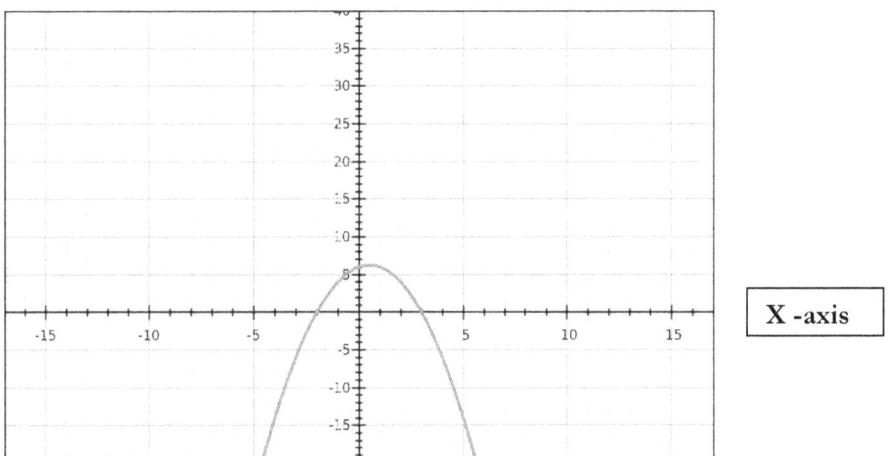

The solution to $-x^2 + x + 8 \geq 2$ is when $-2 \leq x \leq 3$

Example 2:

Solve the inequality $2x^2 + x - 1 > 0$

When it crosses the x-axis we can find the values of x.

That is, $2x^2 + x - 1 = 0$ ➡ $(2x-1)(x+1) = 0$

➡ $x = \frac{1}{2}$ or $x = -1$. Also since the coefficient of x^2 is positive then the parabola will be **U-shaped as shown below**.

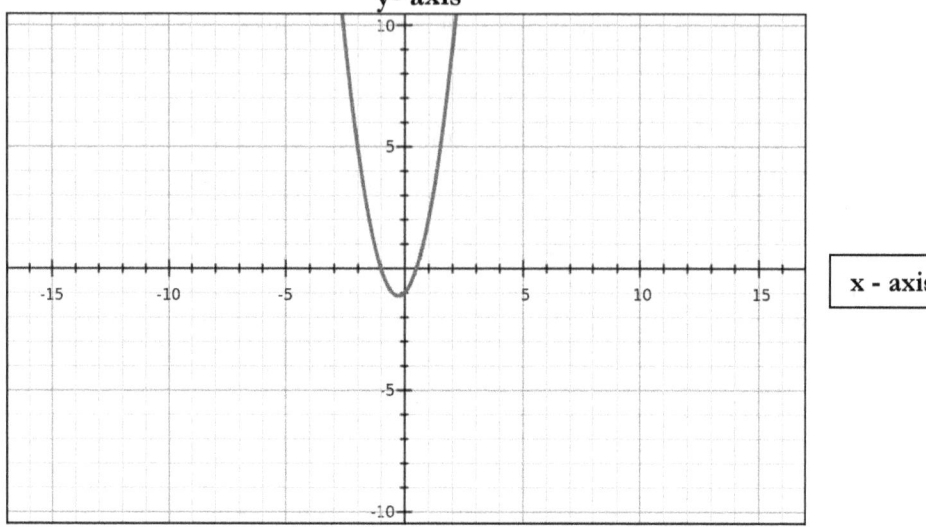

Examining the graph we can see that $2x^2 + x - 1 > 0$ is true when x is less than -1 and when x is greater than $\frac{1}{2}$

That is when $x < -1$ or $x > \frac{1}{2}$

Note: Similar principles have to be applied if you have to solve an quadratic inequality which is < or ≤ rather than > or ≥.

You can best see this visually by actually drawing the graph(s).

Graphs of quadratic equations

Quadratic Equations

These are of the form $f(x) = ax^2 + bx + c$. Note $f(x)$ is called a function of x since y is defined in terms of x.

Consider the example below: Plot the equation $y = 3x^2 - 2x + 1$

First choose suitable values of x say from -3 to + 3 and find the corresponding values of y as shown in the table below:

x	-3	-2	-1	0	1	2	3
$3x^2$	27	12	3	0	3	12	27
-2x	6	4	2	0	-2	-4	-6
+1	1	1	1	1	1	1	1
Y	34	17	6	1	2	9	22

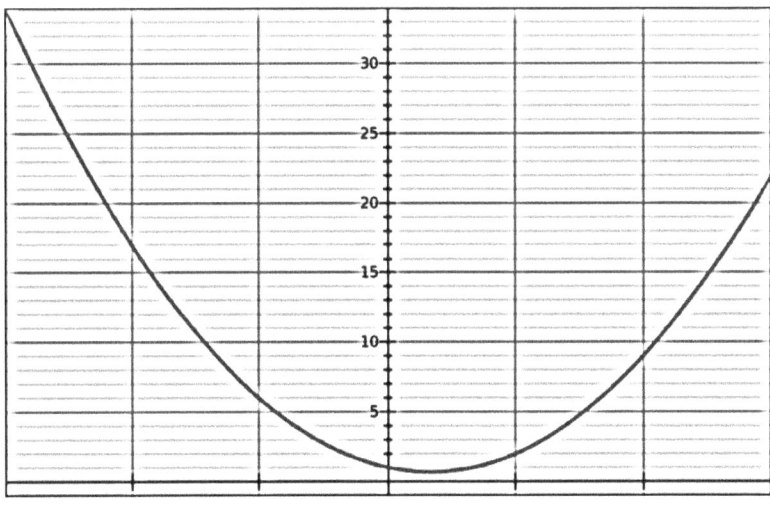

Y - axis

X -axis

Finding the maximum and minimum values of a parabola (that is the turning points of the graph.

For $f(x) = (x + a)^2 + b$ the turning points are at $x = -a$ and $y = b$

In this case the turning point is a minimum

Example: If $f(x) = (x + 2)^2 + 3$, what are the co-ordinates of the turning points and state whether it is a maximum or a minimum. Also state the line of symmetry.

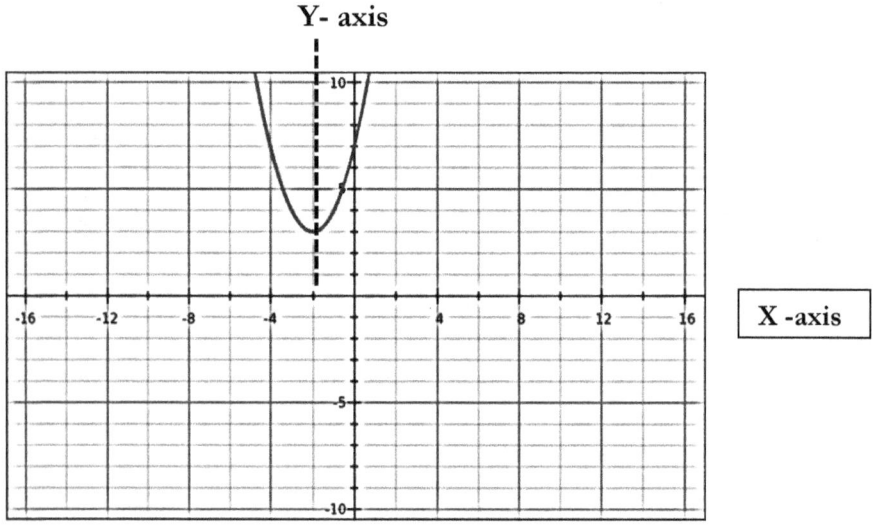

You can see that the graph of $y = (x + 2)^2 + 3$ has a turning point at its minimum -2, 3. The line of symmetry for this function f(x) is at $x = -2$. (Remember f(x) means a function of x, which is the same as $y = (x + 2)^2 + 3$).

Likewise if $f(x) = -(x + a)^2 + b$

Then the turning point is at $x = -a$, $y = b$ and this time is a maximum.

Example 2:

For the equation $y = -(x + 3)^2 + 2$ find the co-ordinates of the turning point and the line of symmetry.

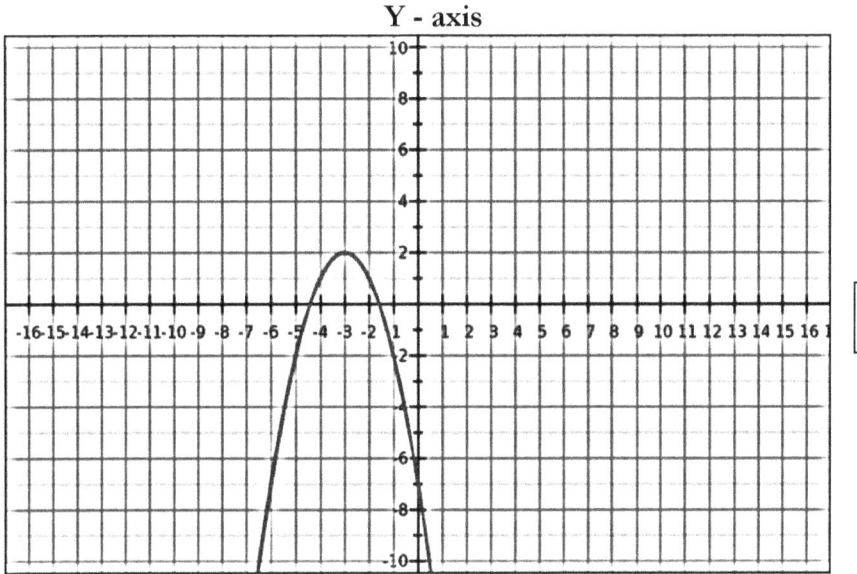

You can see from the graph of $y = -(x + 3)^2 + 2$ above that the turning point is at its maximum where the co-ordinates correspond to (-3, 2). The line of symmetry is at $x = -3$

Note: This means if you are given an equation of a parabola such as $y = x^2 - 2x + 1$, you can re-state it in the form $y = (x + a)^2 + b$, you can then find the co-ordinates of the turning point. In this case the turning point has a minimum value.

Example: Consider the parabola $y = x^2 + 2x + 1$. We can express this as $(x + a)^2 + b = x^2 + 2xa + a^2 + b$. That is the equation of the parabola can be written as $(x + 1)^2 + 0$. Hence the turning point occurs at the point (-1, 0) and it is a minimum.

Cubic equation

Example: Plot the equation $y = x^3 - 1$

x	-3	-2	-1	0	1	2	3
x^3	-27	-8	-1	0	1	8	27
-1	-1	-1	-1	-1	-1	-1	-1
Y	-28	-9	-2	-1	0	7	26

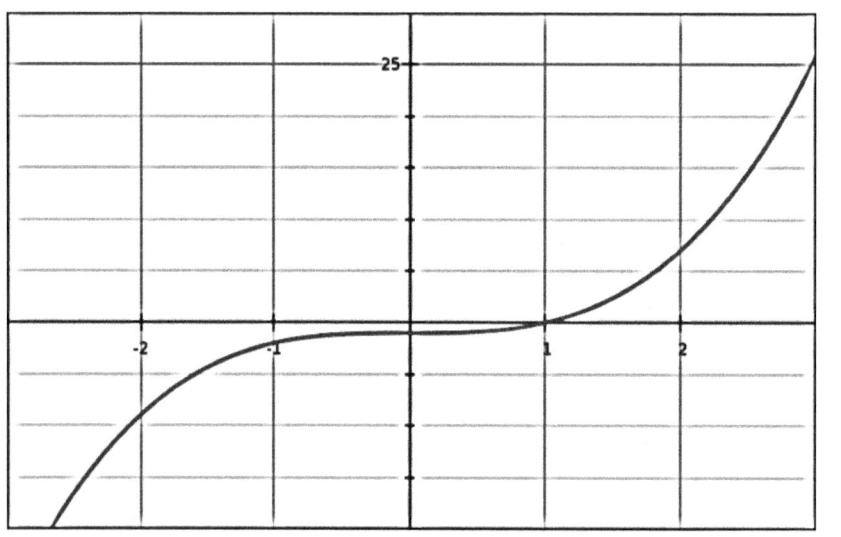

Solving equations using graphical methods

Example

You are given that the quadratic equation $y = x^2 - 4x + 8$ and the linear equation $y = 3x - 2$ intersect at two points A and B. Find the co-ordinates of these two points A and B where the equations intersect.

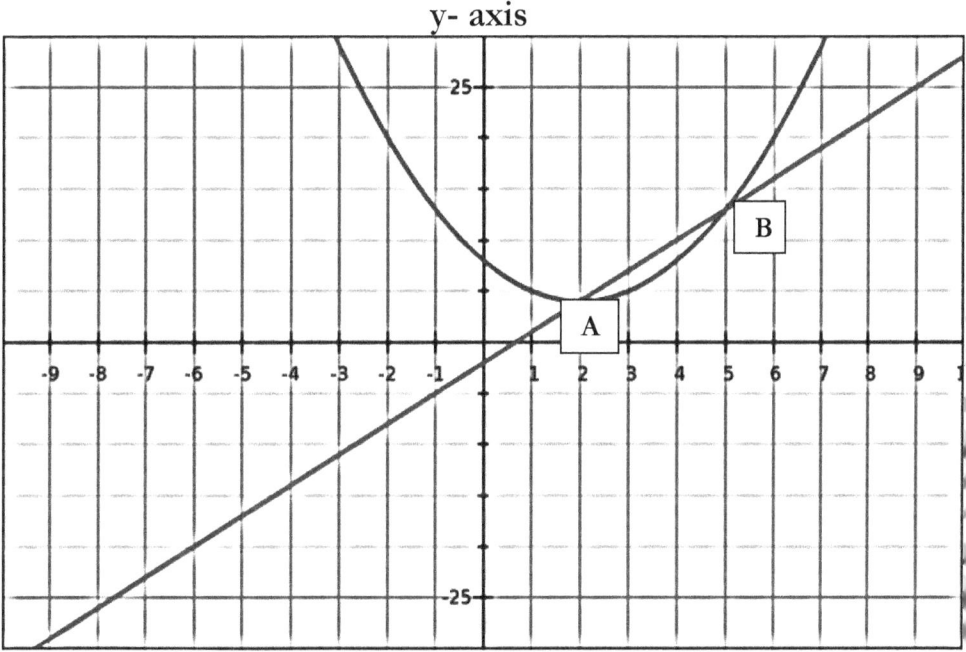

Method: First plot the equations $y = x^2 - 4x + 8$ and the linear equation $y = 3x - 2$ as shown above.

You can see that the co-ordinates of A are (2, 4) and the co-ordinates of B are (5, 13)

Solving equations mathematically when one is linear and the other is quadratic:

The two equations are $y = x^2 - 4x + 8$ and $y = 3x - 2$.

This means $x^2 - 4x + 8 = 3x - 2$

Simplifying this we get $x^2 - 7x + 10 = 0$ ⟹ $(x - 5)(x - 2) = 0$

This means either, $x - 5 = 0$ or $x - 2 = 0$ ⟹ $x = 5$ or 2. We can now find the corresponding values of y by substituting these values in the equation $y = 3x - 2$

When x =5, y = 3×5 – 2 = 13 and when x = 2, y = 3×2 – 2 =4

So the co-ordinates of A = (2, 4) and B = (5, 13)

Transformations of functions

When y = f(x +a) the graph moves 'a' units left. (It is the opposite of what you might expect)

Likewise when y = f(x - a) the graph moves 'a' units to the right

Consider the two graphs below (1) f(x) and (2) f(x) = f(x +2)

(1) $f(x) = x^2 + 2$ and (2) $f(x +2) = (x + 2)^2 + 2$
shown below

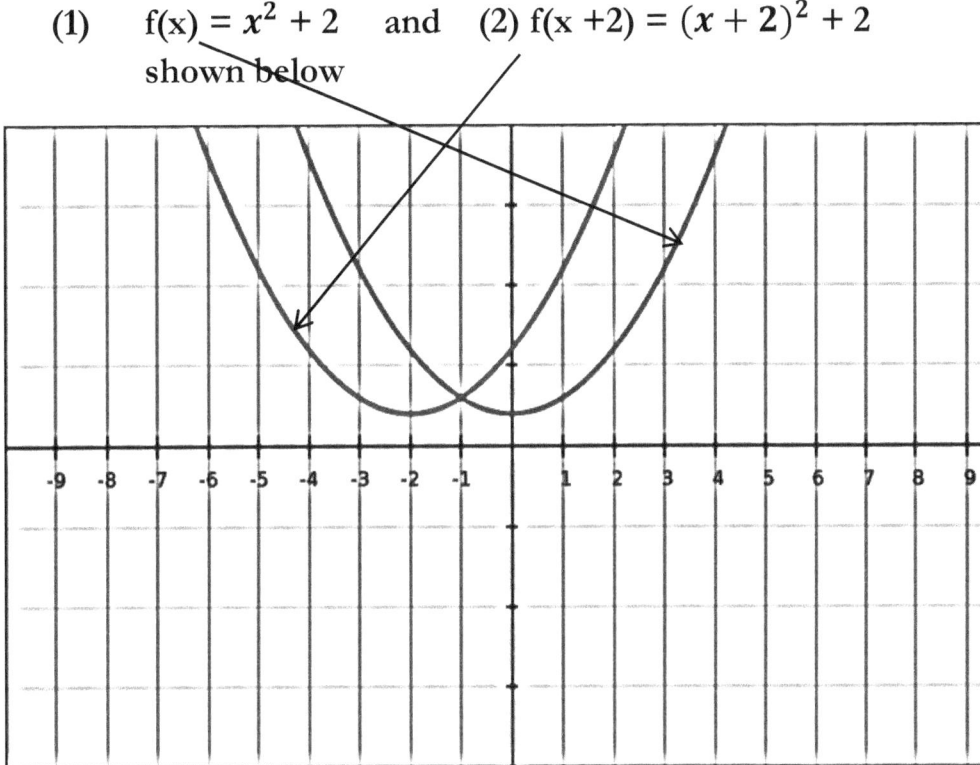

You can see that f(x + 2) has shifted to the left by 2 units (this seems counter-intuitive) but f(x + 2) does shift to the left by 2 units and <u>not</u> to the right.

Other types of transformations involving y = f(x)

It is worth remembering that y = f(x) + a moves up the y-axis by 'a' units and likewise y = f(x) – a moves down the y –axis by 'a' units.

Finally y = k×f(x) or kf(x) simply means the graph of f(x) stretches along the y –axis by a factor of k.

Equation of a Circle

If a circle has radius 'r' and centre (p, q) then its equation is given by:

$$(x - p)^2 + (y - q)^2 = r^2$$

If the circle has its centre at the origin (0, 0) and its radius is '1' then its graph is as shown below. In this case the corresponding equation is $x^2 + y^2 = 1$

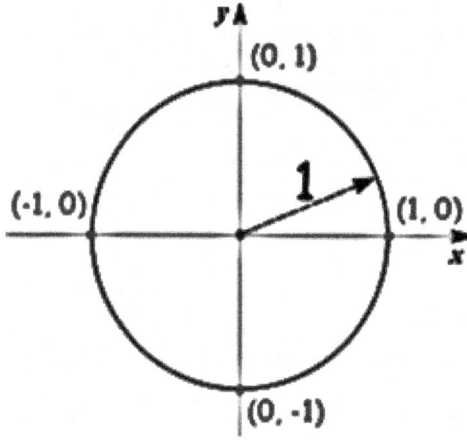

Example 1:

Find the equation of a circle whose radius is 5 and its centre is (2, 3)

<u>Method:</u> Using the formula above the equation of the circle is:

$$(x - 2)^2 + (y - 3)^2 = 5^2$$

We can re-write this as $x^2 - 4x + 4 + y^2 - 6y + 9 = 25$. Simplifying this we get:

$x^2 - 4x + y^2 - 4x - 6y + 13 = 25 \implies x^2 - 4x + y^2 - 4x - 6y - 12 = 0$

Example 2:

Find the centre of a circle and its radius if its equation is $(x - 3)^2 + (y + 4)^2 = 36$. Clearly the centre is 3, -4 and the radius is $\sqrt{36} = 6$
Method: Using the formula for the equation of a circle $(x - p)^2 + (y - q)^2 = r^2$ (where p, q are the centre co-ordinates and r is the radius)

Example 3

Find the centre of a circle and its radius if the equation is $x^2 - 2x + y^2 - 4y - 11 = 0$

Method: Although this looks a bit tricky we need to make this equation similar to: $(x - p)^2 + (y - q)^2 = r^2$ first.

If we expand this we get $x^2 - 2px + p^2 + y^2 - 2yq + q^2 = r^2$

Equating this equation with the one given we get $-2p = -2$ which means $p = 1$

Likewise for the y term we get $-2q = -4$ which means $q = 2$

So the equation is $(x - 1)^2 + (y - 2)^2 = r^2$

Expanding the brackets we get $x^2 - 2x + 1 + y^2 - 4y + 4 - r^2 = 0$

$\implies x^2 - 2x + y^2 - 4y + 5 - 16 = 0$ (we put -16 to adjust to make the equation the same as the original given. $\implies (x - 1)^2 + (y - 2)^2 = 16$
Hence r = 4 and the centre of the circle is 1, 2

Remainder and Factor Theorem

Remainder Theorem

If you have f(x) and you divide it by x – a the remainder will be f(a)

Example 1: Divide $f(x) = 4x^2 - 4x - 1$ by $g(x) = x - 1$

$$\begin{array}{r}
4x, \text{ remainder} - 1 \\
x-1 \overline{\smash{)}\, 4x^2 - 4x - 1} \\
\underline{4x^2 - 4x} \\
0 + 0 - 1
\end{array}$$

Check: Subsitute x= 1 in f(x) we get 4×1×1 - 4×1 – 1 = -1

Example 2: Divide $f(x) = 2x^3 - 3x^2 - 4x - 1$ by x – 3

Method: To find the remainder just work out f(3)

f(3) = 2×3×3×3 - 3×3×3 - 4×3 – 1 = 54 -27 -12 -1 = 14

Hence the remainder is 14.

Factor Theorem

If f(a) = 0, then this implies there is no remainder and x – a is a factor of the polynomial. Conversely, when x – a is a factor of a given polynomial then f(a) = 0

Example 1: Consider $x^2 - x - 2$ and investigate if x – 2 is a factor.

f(2) = 2^2 – 2 – 2 = 4 – 2 – 2 = 0 ⟹ x – 2 is a factor.

Example 2: Consider the cubic polynomial $4x^3 - 4x^2 - x + 1$

Show that 2x – 1, is a factor.

Method: If 2x – 1 is a factor of $4x^3 - 4x^2 - x + 1$, then $f(\frac{1}{2})$ should equal 0.

Let's test this by substituting $x = \frac{1}{2}$ in cubic equation above.

We get $4(\frac{1}{2})^3 - 4(\frac{1}{2})^2 - \frac{1}{2} + 1 = \frac{4}{8} - \frac{4}{4} - \frac{1}{2} + 1 = \frac{1}{2} - 1 - \frac{1}{2} + 1 = 0$. Hence 2x -1 is a factor of $4x^3 - 4x^2 - x + 1$

The factor theorem can be very useful in finding one of the 'roots' of the equation. That is one possible solution to the equation. We can then try and find other factors by dividing the original cubic equation by 2x – 1 and factorise the expression we are left with. Although usually you don't have to divide as you can spot the other factors quickly.

See examples below in solving cubic equations.

Solving cubic equations

Example 1:

Solve the cubic equation $3x^3 + 4x^2 - 3x + 2 = 0$, given that one of its factors is $x + 2$

Step 1: Divide: $3x^3 + 4x^2 - 3x + 2$ by $x + 2$

(We can divide this in the normal long division way as shown below)

$$
\begin{array}{r}
3x^2 - 2x + 1 \\
x+2 \overline{\smash{\big)}\, 3x^3 + 4x^2 - 3x + 2} \\
\underline{3x^3 + 6x^2} \\
-2x^2 - 3x + 2 \\
\underline{-2x^2 - 4x} \\
x + 2 \\
\underline{x + 2} \\
\end{array}
$$

So the cubic equation can be written as $(x + 2)(3x^2 - 2x + 1)$

In other words if one of the factors is $x + 2$, the other factor is $3x^2 - 2x + 1$

Example 2:

Solve the cubic equation $2x^3 + 3x^2 - 3x - 2 = 0$

This time we are not given a factor. In this case it is best to look at the constant term 2 and try $x = 1, -1, 2$ or -2 as possible solutions. (In other words factors of 2) Let's try $x = -1$ as a possible solution. If it is a solution then $f(-1)$ should $= 0$.

Substituting $f(-1)$ we get $2x1^3 + 3x1^2 - 3 \times 1 - 2 = 2 + 3 - 3 - 2 = 0$.

Hence we can re-write $2x^3 + 3x^2 - 3x - 2 = 0$

As $(x - 1)(ax^2 + bx + c) = 0$

Clearly a = 2, since the only way to get $2x^3$ is to multiply x from the first bracket by $2x^2$ in the second bracket. Also we can figure out that c = 2. (Since -1 from the first bracket multiplied by c in the second bracket = -2 which means c = 2.

So far we have $(x - 1)(2x^2 + bx + 2) = 0$. Now multiply the 'x' terms out to get 2x –bx = -3x. This means 5x = bx, hence b =5.

So the factors of $2x^3 + 3x^2 - 3x - 2$ are $(x - 1)(2x^2 + 5x + 2)$

We can now factorise the quadratic expression $2x^2 + 5x + 2$ in the usual way into two brackets. That is $2x^2 + 5x + 2 = (2x +1)((x +2)$.

So finally the cubic equation $2x^3 + 3x^2 - 3x - 2 = 0$ can be written as: $(x - 1)(2x +1)((x +2) =0 \implies$ x = 1 or x =$-\frac{1}{2}$ or x = -2 (Note: We could have also factorised by dividing $2x^3 + 3x^2 - 3x - 2$ by x – 1 using long division shown earlier)

You can also use the graphical method for solving cubic equations as shown below:

Let y = $2x^3 + 3x^2 - 3x - 2$, that is f(x) = $2x^3 + 3x^2 - 3x - 2$. By choosing suitable values of x we can find corresponding values of y and plot the graph. The graph is shown below:

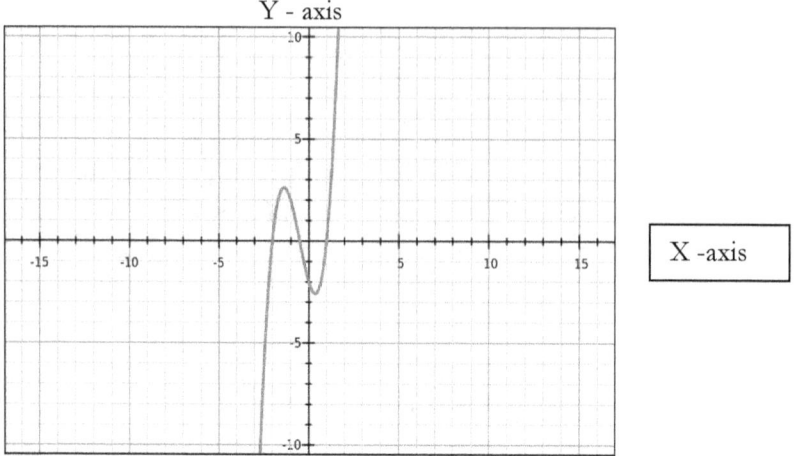

You can see from the graph that the roots are where the curve cuts the x –axis namely at x = -2, -0.5 and 1. Hence the roots of the equation $2x^3 + 3x^2 - 3x - 2$ are when x = -2, or x = -0.5 or x = 1

Linear programming

Basic Principles

Essentially you are trying to find the optimal (best) solution to a given problem with various constraints.

For example a Company may be producing products such as books, computers, shoes, etc. The number or amount of each of these things produced is represented by what is known as a **decision variable** such as x, y, z etc.

Things to note:

(a) There will typically be constraints. These are factors that limit things like the amount of time available, the amount of workers available and so on. Also it is worth noting that constraints can't in general be negative. **For example we can't have -3 workers**!

(b) We need to optimise the function be it profit, cost or other function. So for example when optimising profit we are looking to **maximise** the profit function and when optimising costs we are looking to **minimise** the cost function. We call this the **objective function**

(c) Firstly, we need to find a **<u>feasible</u>** region for optimal solutions of the **objective function** given by all the **constraints**.

(d) Finally, we find that the optimal solutions are found **at one of the vertex points in the feasible region.**
Consider a graphical example below:

Example: Show the feasible region for the optimal solutions given by the following constraints:
(i) $x + y \leq 6$ (ii) $2x - y \geq 2$ (iii) $y > 1$ (Also note that x, y ≥ 0

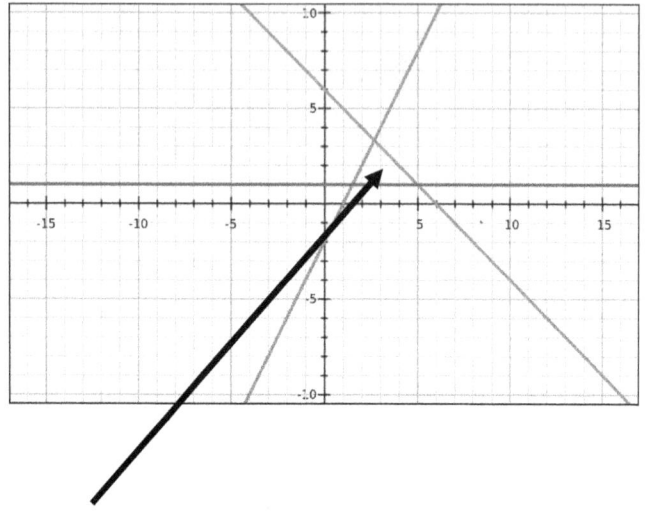

(Feasible Region as shown by the arrow)

Also as we noted earlier in a typical optimisation problem the solution lies at one of the vertices. So we need to find the values of x, y for each of the vertices in the feasible region and substitute these values in the Objective function such as F = ax + by. Depending on the question we may need to find the minimum or maximum value of F to optimise the Objective function.

Now consider a typical exam question

A company sells children's books. It makes x sets of illustrated books and y sets of non-illustrated books for a profit of £5 and £3 each respectively. The constraints are as follows:
(i) $x + y \leq 5$
(ii) $2x - y \leq 7$
(iii) $y \leq 7$

(Assume x, y ≥ 0). Draw a feasible region given the constraints and find the maximum profit for the function F = 5x + 3y

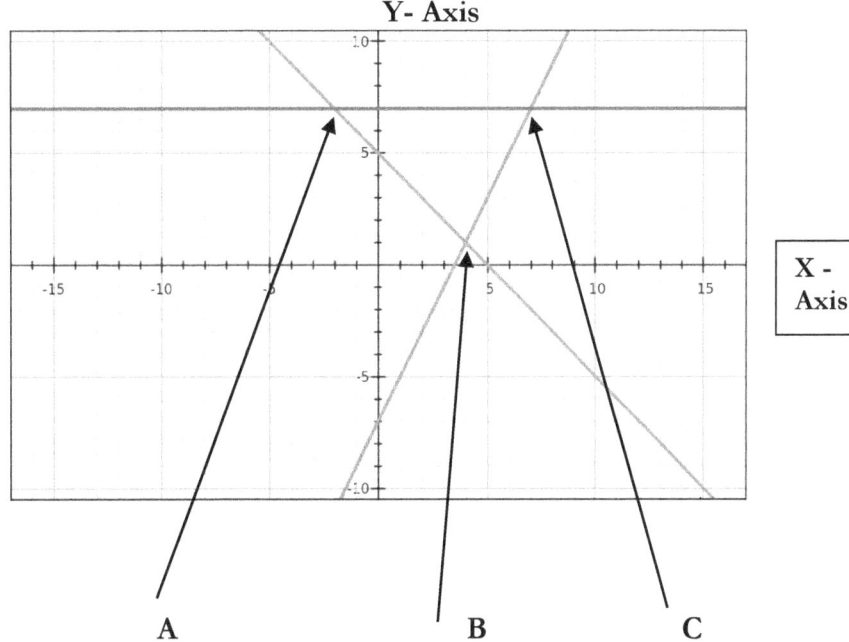

The feasible region is contained in the triangular bit as arrowed by A, B & C

Let's now look at the co-ordinates of each of these vertices, A, B & C

A ={-2, 7}, B = {4, 1} and C = {7, 7}

Clearly the maximum profit for the function F = 5x + 3y is when B ={7, 7}

Substituting (x, y) values for C in F we get F = 5×7 + 3×7 = 35 + 21 = £56

Trig for Right Angled Triangles

Formulae for Right Angled Triangle

In a right-angled triangle you need to know:

The Sine of an angle = the ratio of the Opposite Side to the Hypotenuse

The Cosine of an angle = the ratio of Adjacent Side to Hypotenuse

The Tangent of an angle = the ratio of the Opposite Side to the Adjacent Side

You can also try remembering **SOH-CAH-TOA**

SOH –   SIN (ANGLE) = **OPPOSITE SIDE/HYPOTENUSE**

CAH –   COS (ANGLE) = **ADJACENT SIDE/HYPOTENUSE**

TOA –   TAN (ANGLE) = **OPPOSITE SIDE/ADJACENT**

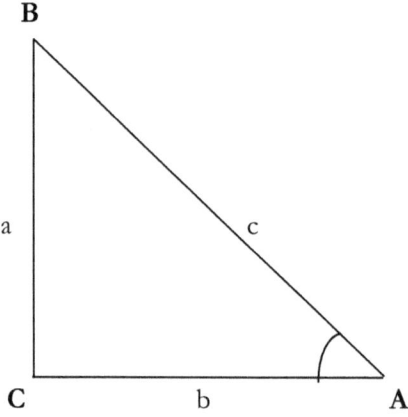

Sin (A) = $\dfrac{a}{c}$

Cos (A) = $\dfrac{b}{c}$

Tan (A) = $\dfrac{a}{b}$

Also, using Pythagoras' theorem you can prove that:

$Sin^2 x + Cos^2 x = 1$

Proof: Consider the triangle below:

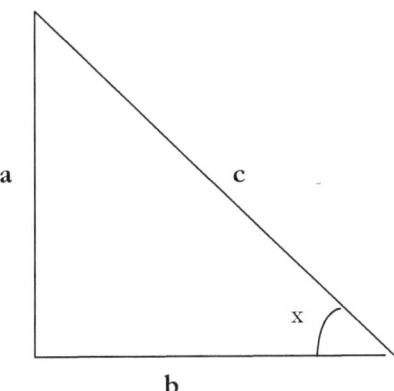

We know that in this triangle

$Sin\ x = \dfrac{a}{c}$ and $Cos\ x = \dfrac{b}{c}$

Hence, $Sin^2 x + Cos^2 x = (\dfrac{a}{c})^2 + (\dfrac{b}{c})^2$

This implies that:

$Sin^2 x + Cos^2 x = \dfrac{(a)^2 + (b)^2}{c^2}$

From Pythagoras' theorem we know that: $a^2 + b^2 = c^2$

Hence, $\text{Sin}^2 x + \text{Cos}^2 x = \dfrac{c^2}{c^2} = 1$

Trig for non- right angled triangles

Formula for a non-right angled triangle are shown below

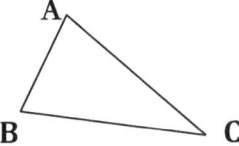

Sine Rule:

$$\frac{a}{SinA} = \frac{b}{SinB} = \frac{c}{SinC}$$

Cosine Rule:

$a^2 = b^2 + c^2 - 2bcCosB$

$b^2 = a^2 + c^2 - 2acCosB$

$c^2 = a^2 + b^2 - 2abCosC$

(Note: Although there are three versions of the formula they all have the same pattern)

Example1: In the triangle below find angle B

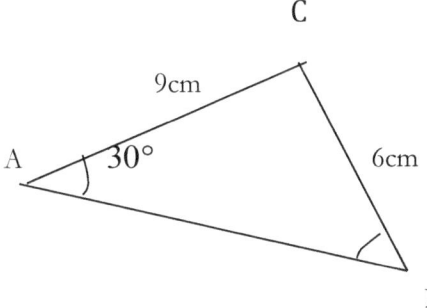

Using the Sine rule: $\dfrac{a}{SinA} = \dfrac{b}{SinB}$ substitute the appropriate known angles and sides in the formula shown

$$\dfrac{6}{Sin30} = \dfrac{9}{SinB}$$

Hence, Sin B = 9 X Sin30/6 = 0.75. So the angle B = 48.6°

Summary for using the sine rule

You can use the sine rule when: You know two angles and a side.

Also note that the area of a non- right angled triangle is given by $\dfrac{1}{2}abSinC$

Using the cosine rule

Example: In the triangle shown find the angle C

```
        B                                    3cm
2.5cm   /\
       /  \
      /    \
     /_____\
    C        A
       4cm
```

Using the Cosine Rule, we can say that $c^2 = a^2 + b^2 - 2abCosC$

Hence, CosC = $(a^2 + b^2 - c^2)/2ab$ = (9 + 16 – 6.25)/24 = 18.75/24 = 0.78125

Hence, C = 38.6°

Summary for using the cosine rule. (1) You know the lengths of 3 sides

Or (2) you know two sides and the angle between them

Graphs of Trig Functions

Consider graphs of y = sin(x) and cos(x)

For y=sin(x) maximum value of y = 1 when x =90° and minimum value of y = - 1 when x= -90°. This cycle repeats every 360°

Also note that when x = 0°, sin(0) =0, and when x = 360°, sin(360) = 0

Similarly for y = cos(x). The maximum value of y = 1, when x = 0° and minimum value of y = -1 when x = 180°. Again this cycle repeats every 360°

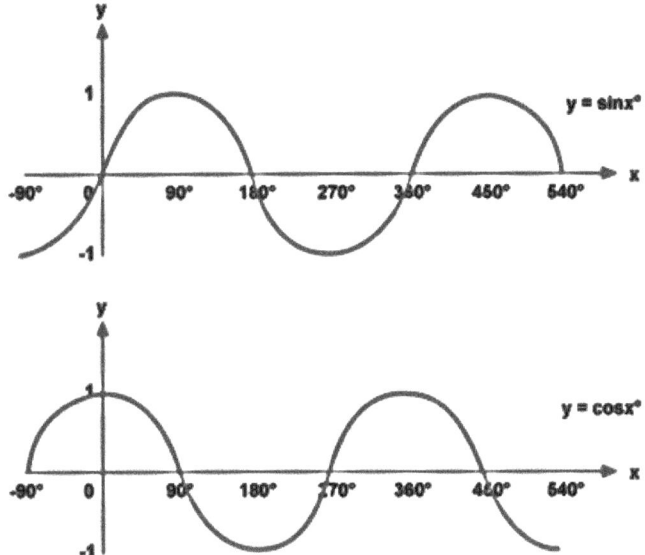

Example of repeating cycles for sin(x)

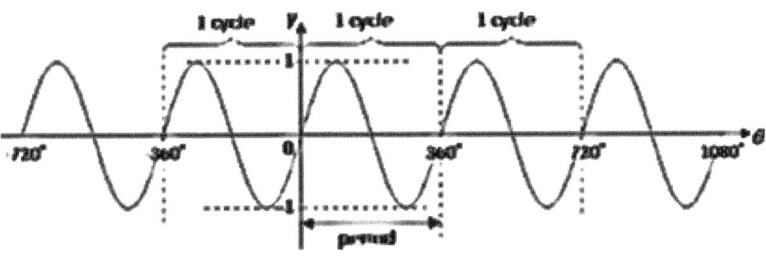

y= tan(x)

The values of tan (x) repeat every 180°. As x approaches 90°, the value of y or tan(x) approaches infinity. The symbol for infinity is ∞.

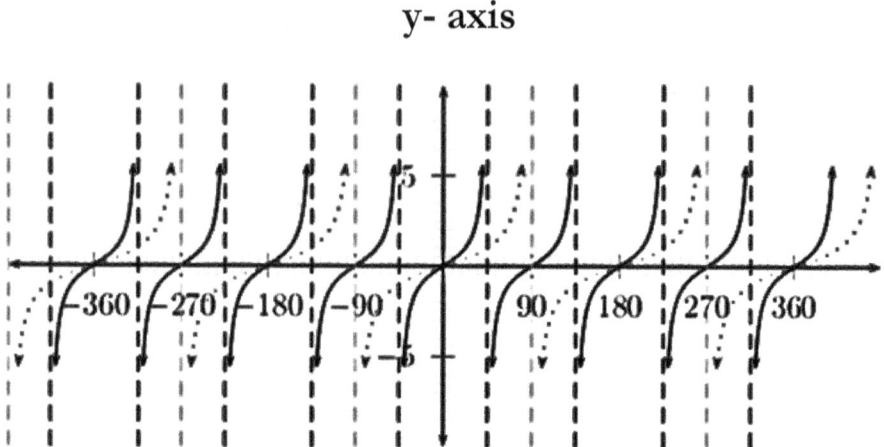

Below is a magnified view of a tangent curve this time of y= tan(θ).

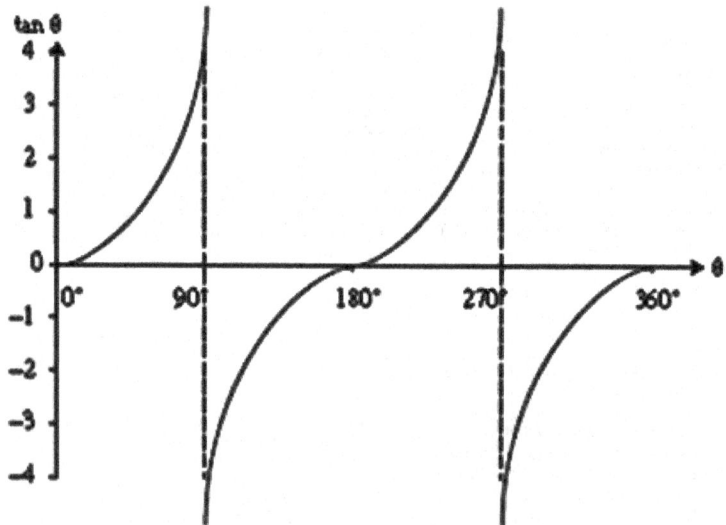

You can see that as θ approaches 90°, tan (θ) tends to infinity

Similarly, as θ approaches -90°, tan (θ) tends to minus infinity

Solving trigonometric equations using graphical methods: Solve the equation sin(x) = 0.5 for values of x between 0 and 360°

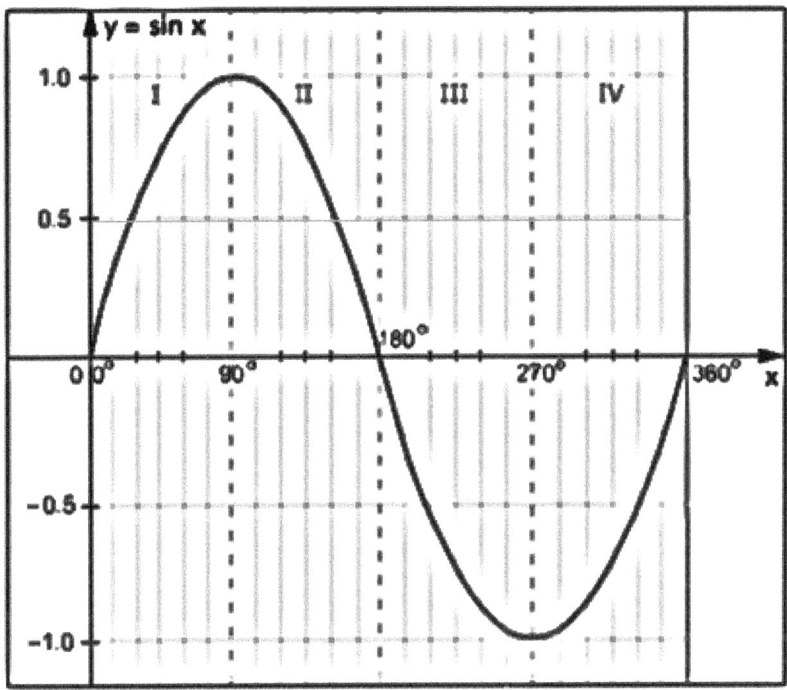

From the graph you can see that when y = 0.5, it meets the curve when x = 30° and 150°

Trig Identities

Rule 1

$sin^2 x + cos^2 x = 1$

Re-arranging we can also deduce that $sin^2 x = 1 - cos^2 x$

Similarly, $cos^2 x = 1 - sin^2 x$

Rule 2

$\text{Tan } x = \frac{sinx}{cosx}$

Example 1: Simplify $sin^2 x + cos^2 x + \frac{sinx}{cosx} - 1$

Method: We know that $sin^2 x + cos^2 x = 1$ and $\frac{sinx}{cosx} = \tan x$

$\implies sin^2 x + cos^2 x + \frac{sinx}{cosx} - 1 = 1 + \tan x - 1 = \tan x$

Example 2: If $\sin x = \frac{1}{2}$, Find the value of cos x in surd form

Method: Using $sin^2 x + cos^2 x = 1$ we can deduce that $\frac{1}{2} \times \frac{1}{2} + cos^2 x = 1$

$\implies \frac{1}{4} + cos^2 x = 1 \implies cos^2 x = \frac{3}{4} \implies \cos x = \frac{\sqrt{3}}{2}$

Example 3: Solve the equation 4sinx + 6 = 9 for values of 0° ≤ x ≤ 90°
(Give your answer to 1 d.p)

Simplifying the equation 4sinx + 6 = 9, we get 4sinx = 3 $\implies \sin x = \frac{3}{4}$
= 0.75. Hence x = 48.6°

Pythagoras' theorem

All you need to remember for this is the formula as shown below.

(Note: This theorem is only true for right angled triangles)

The square of the hypotenuse = the sum of the squares of the other two **sides.**

$$h^2 = a^2 + b^2$$

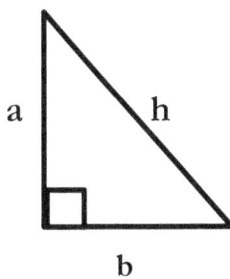

Example 1:

In the triangle below calculate the value of the side b.

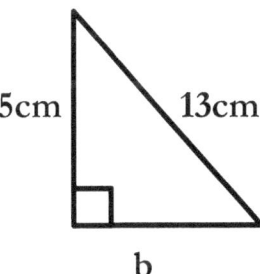

Method: Using Pythagoras' theorem we have $h^2 = a^2 + b^2$

Substituting the values we have $13^2 = 5^2 + b^2$

$\implies 169 = 25 + b^2 \implies 169 - 25 = b^2 \implies 144 = b^2$

Hence $b = \sqrt{144} = 12$

Example 2:

You are given that in the diagram below: HG =12 cm, GE =8cm, BE = 6cm. Find HB. Give your answer to 4 S.F.

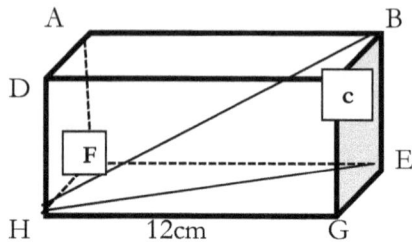

Method: First find HE using Pythagoras' theorem $HE^2 = GE^2 + HG^2$ (Since we know HG & GE)

$\implies$ $HE^2 = 12^2 + 8^2 = 144 + 64 = 208$ $\implies$ $HE = \sqrt{208}$

We can now find HB, since $HB^2 = HE^2 + BE^2$ $\implies$ $HB^2 = 208 + 36 = 244$

$\implies$ $HB = \sqrt{244} = 15.62$ cm

Binomial expansion

Let us first look at Pascal's Triangle. This has the pattern shown below:

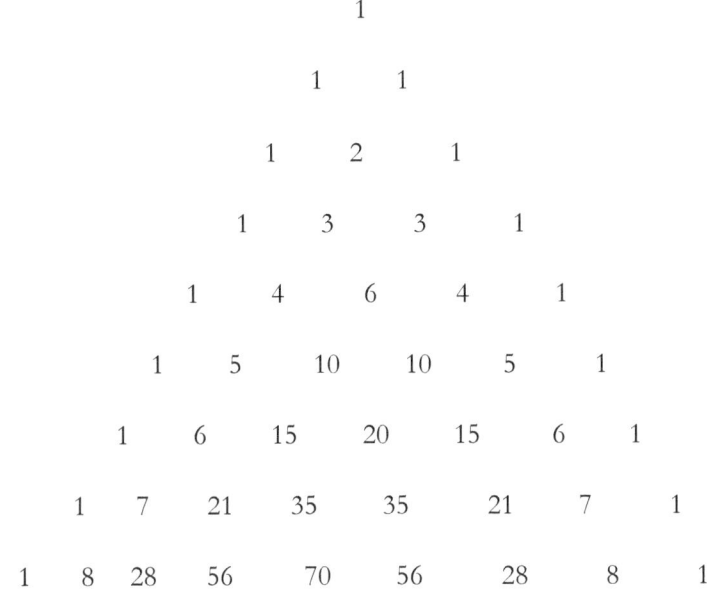

Can you spot the pattern? Each number in the subsequent line is the two numbers from the above line added together!

From this we can arrive at binomial expansions.

Examples:

(1) $(1+x)^0 = 1$
(2) $(1+x)^1 = 1 + x$
(3) $(1+x)^2 = 1 + 2x + x^2$
(4) $(1+x)^3 = 1 + 3x + 3x^2 + x^3$
(5) $(1+x)^4 = 1 + 4x + 6x^2 + 4x^3 + x^4$
(6) $(1+x)^5 = 1 + 5x + 10x^2 + 10x^3 + 5x^4 + x^5$

You can see that the pattern follows Pascal's triangle! Thanks to Pascal a French mathematician.

However, we might need to find a much higher power of the binomial expansion.

Here is a general formula for finding the coefficient of the rth term:
$\frac{n!}{(n-r)!r!}$ also written as $\binom{n}{r}$. (Note $n! = n \times (n-1) \times (n-2) \times (n-3)\ldots\ldots\ldots\ldots \times 2 \times 1$). So $5! = 5 \times 4 \times 3 \times 2 \times 1 = 120$ (the symbol! is called a 'factorial' so n! is 'n' factorial.

Consider the binomial expansion: $(1+x)^8$. We are asked to find the coefficient of the 6th term. In order to find the coefficient of the 6th term in the expansion $(1+x)^8$ we use, $\frac{n!}{(n-r)!r!}$ where n = 8 and r = 6. Hence, $\frac{8!}{(8-6)!6!}$

$\frac{8 \times 7 \times 6 \times 5 \times 4 \times 3 \times 2 \times 1}{(2!) \times 6 \times 5 \times 4 \times 3 \times 2 \times 1} = \frac{8 \times 7}{(2)} = 28$, hence the 6th term is $28x^5$

Binomial Theorem

$(a+b)^2 = a^2 + 2ab + b^2$

$(a+b)^3 = a^3 + 3a^2b + 3ab^2 + b^3$

$(a+b)^4 = a^4 + 4a^3b + 6a^2b^2 + 4ab^3 + b^4$

Binomial Coefficients: $\binom{n}{r} = nC_r = \frac{n!}{(n-r)!r!}$

For example $\binom{4}{2} = \frac{4!}{(4-2)!2!} = \frac{4 \times 3 \times 2 \times 1}{2 \times 1 \times 2 \times 1} = \frac{4 \times 3}{2 \times 1} = 6$

Probability revisited

Probability is defined as the likelihood of an event happening. Probability lies between 0 and 1.

A probability of 0 means that an event will definitely not happen or it is impossible to happen. Likewise a probability of 1 means is certain to happen. Probability is usually expressed as a fraction, a decimal or a percentage.

The probability of an event happening is defined as:

$$\frac{number\ of\ ways\ in\ which\ the\ event\ can\ happen}{total\ number\ of\ outcomes}$$

Also note that the probability of an event **not happening** is **1 – the probability of an event happening**

Notation used: P(A) means probability of event A happening. Hence probability of event A not happening would be 1 – P(A).

Typical examples:

Example 1:

There are 5 red, 6 green and 7 blue beads in a bag.

(1) You pick a bead at random from the bag. What is the probability of picking a red bead? Answer $P(R) = \frac{5}{18}$ (Reason: there are 18 beads altogether, and 5 of them are red, so the chance or probability of picking a red bead is 5 in 18 or $\frac{5}{18}$)

(2) What is the probability of picking a green or blue bead? Answer $P(G\ or\ B) = \frac{13}{18}$

Reason: there are 18 beads altogether, and the number of green and blue beads combined total 13. Hence the probability of picking a green or blue bead is 13 in 18 or $\frac{13}{18}$.

(3) What is the probability of not picking a green bead? Answer: P(not G) = 1 – P(G) = 1 - $\frac{6}{18}$ = $\frac{12}{18}$

A simpler way of doing the same problem is to say that since there are 18 beads altogether and 6 of them are green, then this means that 12 are not green, hence the probability of not picking up a green bead is 12 in 18 that is $\frac{12}{18}$. You could of course simplify $\frac{12}{18}$ to $\frac{2}{3}$ (dividing both the top number 12 and bottom number 18, by 6)

Multiplication law in probability

When you have independent events (that is the outcome of one is not affected by the outcome of the other) then to find the probability of say event A and event B happening we simply multiply the probabilities of A and B together.

Example 1: What is the probability that we will get two sixes when a die is rolled two times?

Method: Probability that we get '6' followed by '6' $= \frac{1}{6} \times \frac{1}{6} = \frac{1}{36}$

Example 2: A fair coin is flipped three times. What is the probability it will turn up 'heads' on all three occasions? Give your answer to 2 decimal places.

Method: Probability that it turns up 'heads' **and** 'heads' **and** 'heads' $= \frac{1}{2} \times \frac{1}{2} \times \frac{1}{2} = \frac{1}{8}$
=0.125 or 0.13 to 2 decimal places. Or you could have calculated it another way i.e. 0.5×0.5×0.5 = 0.25×0.5 =0.125 and then give your answer to two decimal places 0.13 as required.

Example 3:

If a fair die is thrown twice what is the probability of getting a 'six' followed by 'not a six'. Give your answer as a fraction.

Method: P(getting a six) $= \frac{1}{6}$, so the probability of 'not getting a six' $= 1 - \frac{1}{6} = \frac{5}{6}$. Hence the probability that you get a 'six' followed by 'not a six' $= \frac{1}{6} \times \frac{5}{6} = \frac{5}{36}$

Addition law in probability: When two or more events are mutually exclusive (i.e. they cannot occur together), then the probability of A **or** B **or** C happening is simply found by adding the respective probabilities. That is p(A) + p(B) + p(C).

Summary:

(1) Probability lies between 0 and 1 and is usually expressed as a decimal, a fraction or a percentage. The probability of an event can never exceed 1.

(2) When events are independent, to find the probability of A <u>and</u> B occurring together we <u>multiply</u> the probabilities of the respective events. Remember the word <u>'and' is associated with '×' or multiplication</u>.

(3) When events are mutually exclusive the probability of A <u>or</u> B <u>or</u> C happening is found by <u>adding</u> the individual probabilities. Remember the word <u>'or' is associated with '+' or addition</u>.

(4) When working out probabilities consider whether it is 'with' or 'without' replacement

(5) You can generate tree diagrams or sample space diagrams to visualize probabilities and outcomes if it helps you.

Binomial Probability Function

The Binomial Probability function is very useful in working out the probability of r successes out of n trials. (Note in order to use the binomial probability function, **all trials have to be independent and the probability of success p(s) remains constant for each trial**. Like tossing a fair coin 'n' times or throwing a fair 'die' 'n' times. Or tossing an unfair coin or dice with a **known probability** 'n' times. The probability of 'r' successes in 'n' trials is given by: $\binom{n}{r} \times p(s)^r \times p(f)^{n-r}$ (where p(s) is the probability of success and p(f) is the probability of failure.

<u>Notice that the binomial coefficients count the possible outcomes or arrangements of successes and failures</u>.

For example I toss a fair coin 3 times. How many ways are there to arrange heads and tails if it lands 1 head?

Using the binomial coefficient model we saw earlier the number of successes and failures of getting r successes out of n tries was $\binom{n}{r}$. This means the possible arrangements of getting 'heads' and 'tails' if I throw a coin three times with one 'head' is $\binom{3}{1} = \frac{3!}{(3-1)!1!} = \frac{3 \times 2 \times 1}{(2 \times 1) \times 1} = 3$ possible arrangements.

(You can do this manually of course: HTT, THT, TTH. You can see that there are 3 possible outcomes! The binomial probability distribution simplifies this for us as the number of throws (n) becomes bigger.

Example 1: I throw a fair dice 4 times. Find the probability of rolling 2 sixes.

We know that p(s) is the probability of success is for getting a 'six' is $\frac{1}{6}$ and the probability of not getting a 'six' is $\frac{5}{6}$. Using the binomial probability distribution we can work out the probability of two 'successes' out of 'four' throws is given by $\binom{4}{2} \times (\frac{1}{6})^2 \times (\frac{5}{6})^{4-2} =$
$\frac{4!}{(4-2)!2!} x (\frac{1}{6})^2 x (\frac{5}{6})^2 = 6 \times (\frac{1}{36}) x (\frac{25}{36}) = (\frac{1}{6}) x (\frac{25}{36}) = \frac{25}{216} = 0.116$ to 3 decimal places

Example 2: I toss an unbiased coin 6 times. What is the probability of getting 4 heads out of 6 independent throws?

Method: Using the binomial probability distribution the probability of getting 4 heads out of 6 throws is $\binom{6}{4} \times (\frac{1}{2})^4 \times (\frac{1}{2})^{6-4} = \frac{6!}{(6-4)!4!} x (\frac{1}{2})^4 x (\frac{1}{2})^2$

$= \frac{6 \times 5 \times 4 \times 3 \times 2 \times 1}{(2 \times 1) \times (4 \times 3 \times 2 \times 1)} \times (\frac{1}{2})^4 x (\frac{1}{2})^2 = \frac{6 \times 5}{(2 \times 1)} \times (\frac{1}{16}) \times (\frac{1}{4}) = 15 \times \frac{1}{64} = \frac{15}{64}$

Example 3 18 people are given a drug for a certain illness. The probability that an individual has a side effect is 0.12.

(a) Find that no one in this group of 18 has a side effect
(b) Find the probability that in this group 1 individual has a side effect.

(a) **Method:** The probability of having a side effect is 0.12

$\implies$ Probability of not having a side effect = 0.88

$\implies$ The probability that no one in this group has a side-effect is equal to $0.88^{18} = 0.1001588\ldots$ or **0.1002 to 4 decimal places**

(b) **Method:** Using the binomial probability function the probability that 1 individual in this group of 18 has a side-effect =
$\binom{18}{1} \times 0.12^1 \times 0.88^{17} = 0.245844\ldots$ = **0.2458 to 4 decimal places.**

Example 5 A biased coin has a probability of 0.4 in landing heads when tossed. The coin is tossed 5 times. Find the probability that it lands heads at least twice.

Method: The probability that it lands heads at least twice

$= 1 - \{p(0) + p(1)\} = 1 - p(0) - p(1)$

$= 1 - \binom{5}{0} \times (0.4)^0 \times (0.6)^5 - \binom{5}{1} \times (0.4)^1 \times (0.6)^4$

$= 1 - (0.6)^5 - 5 \times 0.4 \times (0.6)^4 = 1 - 0.07776 - 0.2592$

$= \mathbf{0.6630 \text{ to 4 decimal places}}$

(Note: p(0) in this case means probability of no heads and p(1) means the probability of getting 1 head in the 5 throws)

Differentiation

If y = f(x) then differentiating this function is finding the rate of change of y with respect to x. **It is also the gradient at a point on the curve.** It is represented by $\frac{dy}{dx}$.

If $y=x^n$, then $\frac{dy}{dx} = nx^{n-1}$

Example 1: Differentiate the function $y = 2x^3 + 3x + 7$, and find its gradient when x = 3

Method: First differentiate this function

$\frac{dy}{dx} = 6x^2 + 3 + 0 = 6x^2 + 3$, hence the gradient at x = 3 is 6×3×3 + 3 = 18×3 + 3 = 54 + 3 = 57.

(Note: when differentiating a constant e.g. any number c we get 0)

Hence the gradient $\frac{dy}{dx}$ when x= 3 for this function is 57.

Example 2:

You are told that a vehicle travels s metres in t seconds. The formula is given by the equation $s = 4t^2$, find the speed of this vehicle when t = 3 seconds.

Method: $\frac{ds}{dt}$ represents the rate of change of distance with respect to time. This in fact is the speed.

So differentiating, $s = 4t^2$ we get $\frac{ds}{dt} = 8t$. This means the speed of this vehicle at 3 seconds is 8×3 = 24 metres/second.

Stationary points, minima and maxima

For a curve the stationary point occurs when $\frac{dy}{dx} = 0$. (The gradient = 0). To establish whether it is a maximum or minimum we need to find the second derivative. If the second derivative namely, if $\frac{d^2y}{dx^2} > 0$, then this stationary point is a minimum. Also if $\frac{d^2y}{dx^2} < 0$, then the stationary point is a maximum.

Example 3:

You are given that the equation of a curve is $y = 3x^3 - 2x^2 + 2x + 6$

    (a) Find the gradient of this curve when x = 1
    (b) Find the equation of the tangent to this curve at this point
    (c) Find the equation of the normal at this point (x =1)

(a) **Method:** Gradient = $\frac{dy}{dx}$ = $9x^2 - 4x + 2$

$\Rightarrow$ at x =1 the gradient ($\frac{dy}{dx}$)= 9×1×1 - 4×1 +2 = 9 – 4 +2 = 7

(b) **Method:** When x =1 then the value of y can be found by substituting for x in is y = $3x^3 - 2x^2 + 2x + 6$ $\Rightarrow$ y =3× 1^3 -2× 1^2 + 2×1 + 6 = 3 -2 +2 +6 =9 $\Rightarrow$ y = 9. Hence when x = 1, y = 9 and the gradient is 7

Since the equation of a straight line can be found by using y – y1 = m(x –x1)

By substituting the appropriate values we get y – 9 = 7(x – 1)

$\Rightarrow$ y – 9 = 7x – 7 $\Rightarrow$ y = 7x + 2

(c) **Method**: To find the equation of the normal at this point we first need to find its gradient. Using the fact that m1m2 =- 1 **(Note if m1 is the gradient of a tangent at a given point and m2 the gradient of the normal at that point then m1×m2 = -1)**

We can deduce that 7×m2 = -1 $\Rightarrow$ m2 = $-\frac{1}{7}$

$\Rightarrow$ The equation of the normal at this point is y – 9 = $-\frac{1}{7}$(x – 1)

Simplifying we get 7y – 63 = -x + 1 $\Rightarrow$ 7y = 64 – x

Or $y = \frac{1}{7}(64 - x)$

Integration

This is the reverse of differentiation. Essentially when you integrate a function you find the original function you started with. There are two types of integrals 'indefinite' and 'definite'. Definite integrals are used to find an area under a curve.

Rule for integrating: $\int x^n dx = \frac{x^{n+1}}{n+1} + c$ (where c is the constant of integration). This only applies to indefinite integrals.

Example 1: Find $\int x^2 dx$

Method: Using the formula for indefinite integrals we get $\int x^2 dx = \frac{x^{2+1}}{2+1} + c = \frac{x^3}{3} + c$

Example 2: Find $\int (x^4 + 3x) dx$

Method: Using the above method we get $\int (x^4 + 3x) dx = \frac{x^{4+1}}{4+1} + 3\frac{x^{1+1}}{1+1} + c = \frac{x^5}{5} + 3\frac{x^2}{2} + c$

Example 3:

Find $\int (x^2 + 2x + 5) dx$

Method: Using the method above we get: $\int (x^2 + 2x + 5) dx = \frac{x^3}{3} + \frac{2x^2}{2} + 5x + c = \frac{x^3}{3} + x^2 + 5x + c =$

(Note: You can write '5' as $5x^0$ so $\int 5 dx = \int 5x^0 dx = 5x^{0+1} = 5x$)

Example 4: Find $\int (x^{-2} + 5x) dx$

Method: Applying the formula we find that $\int (x^{-2} + 5x) dx = \frac{x^{-1}}{-1} + \frac{5x^2}{2} + c$
$= -\frac{1}{x} + \frac{5x^2}{2} + c$

Example 5: Find $\int (x^{\frac{1}{3}} + 2x)dx$

Applying the formula we get $\int (x^{\frac{1}{3}} + 2x)dx = x^{\frac{4}{3}} + \frac{2x^2}{2} + c = x^{\frac{4}{3}} + x^2 + c$

Definite Integrals

These can be used to find the area under a curve or a function. The way to do this is the same as integrating normally first, then substitute the limits and subtract as shown below in the example.

Example: Find the value of $\int_2^3 2x\ dx$

Method: Integrate normally, substitute the values and subtract as shown below:

$\int_2^3 2x\ dx = 2 \times \frac{x^2}{2} = [x^2]_2^3 = 3^2 - 2^2 = 9 - 4 = 5$

Some areas are easy to find but some can be a bit tricky!

Easy example:

The graph below shows equation y = x^3 − 1. Find the area bounded by the curve and x= 2 and x = 1 as shown by the shaded area.

Y -axis

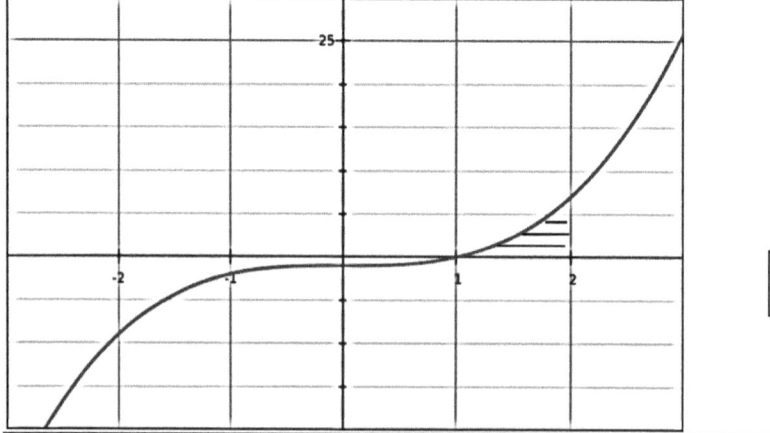

X -axis

The area required can be worked out by the definite integral below:

$$\int_1^2 (x^3 - 1)dx = \left[\frac{x^4}{4} - x\right]_1^2 = \{(\frac{16}{4} - 2) - (\frac{1}{4} - 1)\} = 2 - (-\frac{3}{4}) = 2 + \frac{3}{4} = 2\frac{3}{4}$$

So the area of the equation $y = x^3 - 1$ between 2 and 1 is $2\frac{3}{4}$

Harder examples: Areas enclosed by say a curve and a line as shown below:

Example 2: You are given that $f(x) = x^2 + 3x$ and $g(x) = 5 - x$

The graph of the two equations is shown below. Find the area enclosed above the x-axis and between 0 and 5

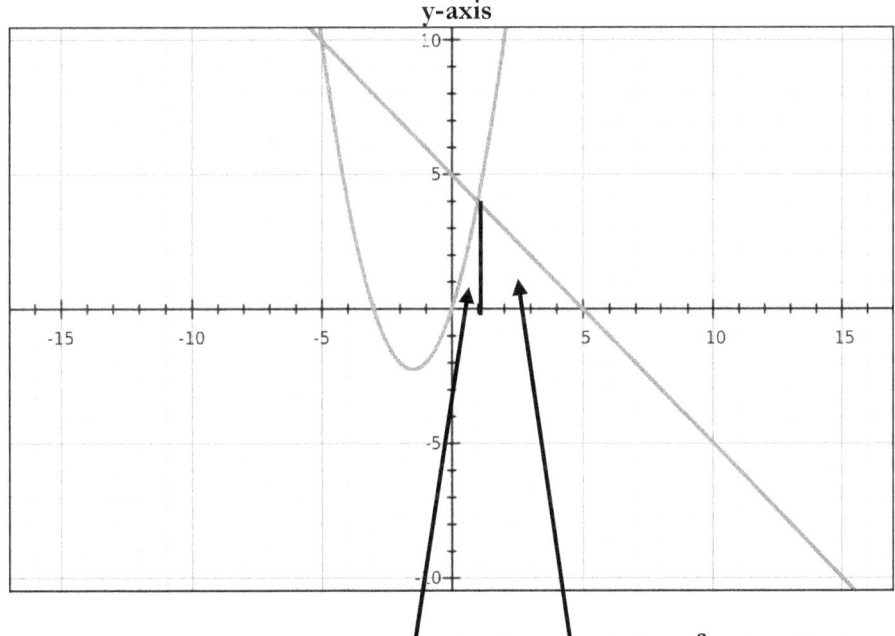

Clearly we want the area between 0 & 1 for the curve $f(x) = x^2 + 3x$ added to the area between 1 and 5 for the line $y = 5 - x$. That is area A + area B

Area A = $\int_0^1 (x^2 + 3x)\,dx = \left[\frac{x^2}{3} + \frac{3}{2}x^2\right]_0^1 = \frac{1}{3} + \frac{3}{2} = \frac{2+9}{6} = \frac{11}{6} = 1\frac{5}{6}$

Area B = $\int_1^5 5 - x\,dx = \left[5x - \frac{x^2}{2}\right]_1^5 = (25 - 12\frac{1}{2}) - (5 - \frac{1}{2}) = 12\frac{1}{2} - 4\frac{1}{2} = 8$

$\Rightarrow$ A + B = $1\frac{5}{6} + 8 = 9\frac{5}{6}$

$\Rightarrow$ Area under the curve and enclosed by the line is = $9\frac{5}{6}$ square units

Finding areas below the x- axis

If you integrate the normal way for finding areas below the x-axis you will end up with a negative area. Don't worry about this. Common sense tells you that you can't have a negative area so you simply take the 'absolute' value or put more simply just take the positive value as the answer.

Example: Find the area enclosed by the curve and the x –axis between -2 and 0 as shown by the shaded area.

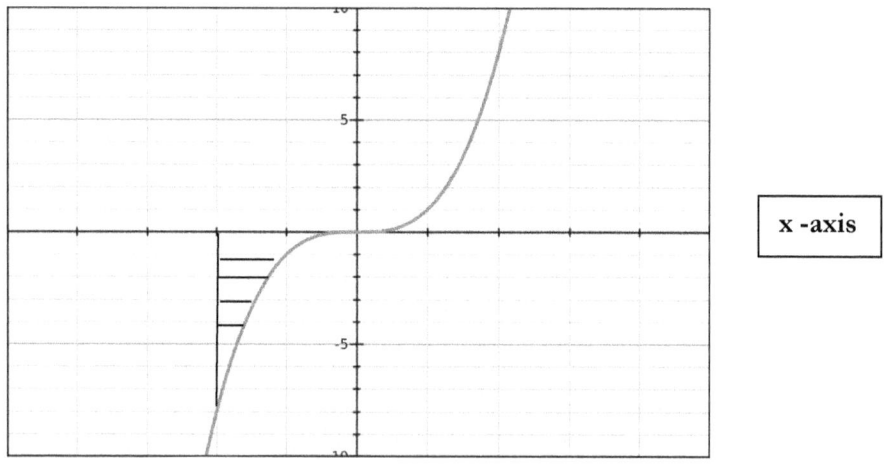

Area required is the absolute value of $\int_{-2}^{0} x^3 dx = \left[\frac{x^4}{4}\right]_{-2}^{0}$

Or $= \left[\frac{x^4}{4}\right]_{0}^{-2} = \frac{16}{4} - 0 = 4$ square units

(Note: we change the order of 0 and -2, since we want a positive value)

Finding areas enclosed between curves

Example: Find the area enclosed between y= $x^2 + 3$

And y = 21 - x^2. The two curves are shown below:

(Note the two curves intersect at x = -3 and x = 3)

Since if y= $x^2 + 3$ and y = 21 - x^2 ⟹ $x^2 + 3 = 21 - x^2$

⟹ $2x^2 = 18$ ⟹ $x^2 = 9$ ⟹ x = $\sqrt{9}$ ⟹ x = 3 or – 3

We have to integrate between -3 and + 3 and subtract one from the other as shown in the working out.

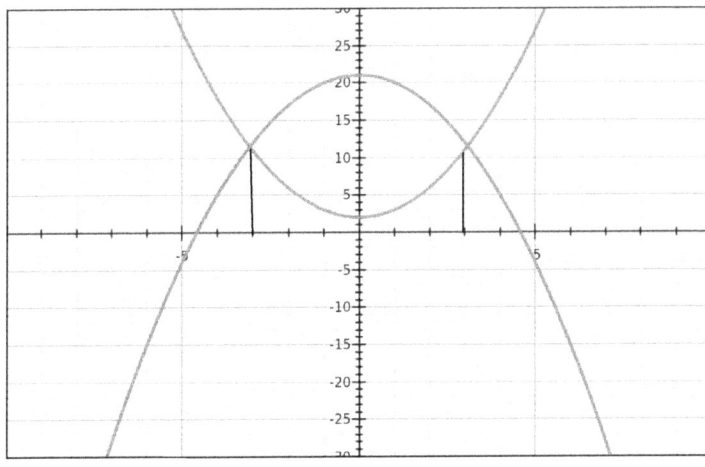

Area required = Area of the inverted U-shaped curve minus the area of the U-shaped curve between the intervals shown x = -3 and x =3.

Area under inverted U-shaped curve between x =-3 and x =3 is

$\int_{-3}^{3}(21 - x^2)\,dx = \left[21x + \frac{x^3}{3}\right]_{-3}^{3} = (63 - 9) - (-63 - (-9))$

$= 54 + 63 - 9 = 108$ square units

Similarly the area under the U-shaped curve between these intervals is:

$\int_{-3}^{3}(x^2 + 3)\,dx = \left[\frac{x^2}{3} + 3x\right]_{-3}^{3} = (9 + 9) - (-9 + (-9))$

$= 18 + 9 + 9 = 36$ square units.

Hence the area between the curves in the interval x =-3 and x =3 is:

108 - 36 = 72 square units.

Kinematics

If you have done or are doing Physics then you have probably come across the equations below. These equations involve objects travelling in a straight line. You need to remember these equations!

(1) $v = u + at$

(2) $s = ut + \frac{1}{2}at^2$

(3) $v^2 = u^2 + 2as$

Where v = final velocity, u = initial velocity, s= distance travelled

a = acceleration and t = time taken.

Units for speed are usually ms^{-1}, distance is in metres and acceleration in ms^{-2}.

We can also derive that $s = \frac{t(u+v)}{2}$ from (1) and (2)

From (1) $v = u + at$, we can deduce that $a = \frac{v-u}{t}$

Now substitute for a in equation $s = ut + \frac{1}{2}at^2$

$\Rightarrow s = ut + \frac{(v-u)t^2}{2t}$ $\Rightarrow s = ut + \frac{(v-u)t}{2}$

$\Rightarrow s = \frac{2ut + (v-u)t}{2}$ $\Rightarrow s = \frac{2ut + vt - ut}{2}$

$\Rightarrow s = \frac{ut + vt}{2}$ $\Rightarrow s = \frac{t(u+v)}{2}$

(This formula, $s = \frac{t(u+v)}{2}$ is also worth remembering)

Consider the following examples:

Example 1: A cyclist travels in a straight line and covers a distance of 100m with an initial velocity of $5ms^{-1}$. Assuming its accelerating at $1\ ms^{-2}$, find the final velocity after covering this distance.

Method: Using the formula $v^2 = u^2 + 2as$ and substituting the appropriate values we find that

$v^2 = 5^2 + 2 \times 1 \times 100 \implies v^2 = 25 + 200 \implies v^2 = 225 \implies v = \sqrt{225} = 15ms^{-1}$. Hence the final velocity is $15ms^{-1}$

Example 2: An object is travelling at $20ms^{-1}$. It starts to accelerate and reaches a final velocity of $30ms^{-1}$ after 10 seconds has elapsed. Find the acceleration.

Method: Using the formula $v = u + at$ and substituting the appropriate values we find that

$30 = 20 + a \times 10 \implies 10 = 10a \implies a = 1ms^{-2}$

Example 3: An object is dropped from a height of 20 metres. (The initial velocity is 0.) The acceleration due to gravity is $g = 9.8ms^{-2}$. Calculate the time taken when it reaches the ground.

Method: Using the formula $s = ut + \frac{1}{2}at^2$ *and* making the appropriate substitutions we get:

$20 = 0 + \frac{1}{2} \times 9.8 \times t^2 \implies 40 = 9.8 \times t^2 \implies t^2 = \frac{40}{9.8} \implies t = \sqrt{\frac{40}{9.8}} = 2.02$ seconds to two decimal places.

Things to note:

(1) A velocity time graph which is linear (a straight line) means the acceleration is constant
(2) The steeper the gradient (slope) of this line the faster the object is moving
(3) If the velocity time graph is a curve this means its acceleration is changing and to find its acceleration at a given point in time we need to find its gradient.

Differentiation and Integration in Kinematics

Using differentiation and Integration in Kinematics

Example 1: Assume that a particle travels s metres after t seconds where $s = 3t^3 - 2t^2$

Find (i) the velocity after two seconds (ii) Find the acceleration after 3 seconds

Method: (i) Since $v = \frac{ds}{dt}$. Since we are given that $s = 3t^3 - 2t^2$ ⟹ $v = \frac{ds}{dt} = 9t^2 - 4t$

Hence when t = 2 seconds $\frac{ds}{dt} = 9 \times 2 \times 2 - 4 \times 2 = 36 - 8 = 28$ ms^{-1}

(ii) To find the acceleration we simply differentiate the velocity. Above we found that $v = 9t^2 - 4t$

⟹ $a = \frac{dv}{dt} = 18t - 4$ ⟹ after 3 seconds the acceleration (a) = $18 \times 3 - 4 = 50$ ms^{-2}

Example 2: A car starts from rest and accelerates for 5 seconds before reaching a constant velocity. The equation for this time interval is given by $v = 12t - \frac{t^2}{2}$ ms^{-1}

(a) Write down the expression for the car's acceleration during the first 5 seconds

(b) Find the distance the car travels in 5 seconds

Method: (a) since $v = 12t - \dfrac{t^2}{2}$ ms^{-1} ⟹ $a = \dfrac{dv}{dt} = 12 - \dfrac{2t}{2} = 12 - t$

(c) To find the distance travelled in 5 seconds we need to re-write $v = 12t - \dfrac{t^2}{2}$

⟹ $\dfrac{ds}{dt} = 12t - \dfrac{t^2}{2}$ ⟹ $s = \int_0^5 (12t - \dfrac{t^2}{2}) = \left[\dfrac{12t^2}{2} + \dfrac{t^3}{6}\right]_0^5 =$

$= 6\times 25 - \dfrac{125}{6} = (150 - 20\dfrac{5}{6}) - (0)$

$= 129\dfrac{1}{6}$ m ⟹ The distance travelled in 5 seconds is $129\dfrac{1}{6}$ m

Formula Sheet

In any triangle ABC

Cosine Rule: $c^2 = a^2 + b^2 - 2ab\cos C$

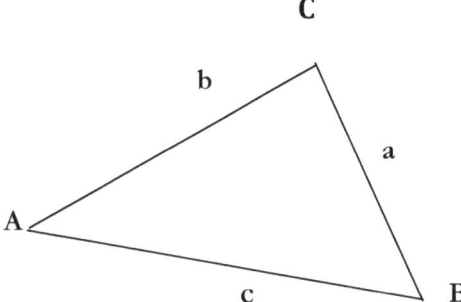

Binomial Expansion

Where n is a positive integer

$$(a + b)^n = \binom{n}{1} a^{n-1}b + \binom{n}{2} a^{n-2}b^2 + \ldots\ldots\binom{n}{r} a^{n-r}b^r \ldots\ldots b^n$$

Where $\binom{n}{r} = nC_r = \dfrac{n!}{(n-r)!\,r!}$

Practice Test 1   Section A   Total 52 Marks

(1)(a) Find the equation of the tangent to the curve $y = 2x^2 - 2x + 5$
at $x = 1$, $y = 5$ 3 marks

  (b) Find the equation of the normal at the same point   3 marks

(2) Solve the inequality $3(x + 7) \geq 5x + 3$ 3 marks

(3) If $\sin(x) = \dfrac{2}{3}$

  (a) find the values of x where $0 \leq x \leq 180°$ 3 marks

  (b) Prove that $\tan(x) = \dfrac{2}{\sqrt{5}}$ 4 marks

(4) Two points on a straight line P and Q have co=ordinates P(1, 3) and Q(3, 7).

  (a) Find the mid-point of PQ   1 mark

  (b) Find the equation of the straight line that goes through P and Q   3 marks

  (c) What is the gradient of the line that is perpendicular to PQ
                                                   2 marks

(5) A curve has the equation $f(x) = 3x^2 - 2x + 5$

  (a) Find the co-ordinates of the stationary point   4 marks
      Method:

  (b) Work out whether it is a maximum or minimum   3 marks

(6) (a) Simplify the expression $y^{\frac{2}{5}}(y^{\frac{-2}{3}} + y^{\frac{1}{3}})$ 3 marks

(b) Simplify the expression below: 4 marks

$$\frac{2}{x-3} - \frac{3}{x+3} + \frac{1}{x^2-9}$$

(7) Solve the quadratic equation $6x^2 + x - 1 = 0$ 3 marks

(8) The equation of a circle is given by $(x-3)^2 + (y-4)^2 = 49$

(a) Find the centre of the circle as well as the length of its diameter   3 marks

(b) Sketch the circle on a graph   2 marks

(9) Find the intersection points where the line $y = x + 2$ meets the curve $y = x^2 - x + 1$ 4 marks

(10) Given two equations in kinematics (i) $v = u + at$ and (ii) $s = ut + \frac{1}{2}at^2$. Prove that $s = \frac{t(u+v)}{2}$ 4 marks

Section B   48 marks

(11) (a) Using the remainder theorem show that when $f(x) = 2x^3 - x^2 - 18x + 12$ is divided by $x - 3$ the remainder is 3.   3 marks

(b) Prove that the remainder is 3, by dividing $f(x)$ by $x - 3$ using long division   4 marks

(c) If $g(x) = (x - 1)(2x^3 - x^2 - 18x + 12)$ work out $g(-2)$ 3 marks

(d) Evaluate $fg(1)$ 2 marks

(12) A biased coin has a probability of 0.42 of landing heads when tossed. This coin is tossed 15 times.

(a) Find that it does not land heads in all the 15 throws. Give your answer to 5 decimal places.   4 marks

(b) Find the probability that it lands heads at least twice.

    8 marks

(13) A straight line given by the equation $y = 2x + 2$ intersects a curve $y = 2x^2 + 3x - 1$ at two points P and Q.

(a) Find the co-ordinates of these two points.   5 marks

(b) Find the equation of the line that is parallel to $y = 2x + 3$ and goes through $(-2, -3)$ 3 marks

(c) Work out whether the curve $y = 2x^2 + 3x - 1$ has a maximum or minimum turning point.   4 marks

(14) A particle moves distance s metres in time t seconds given by the equation $S = \dfrac{3t^3}{2} - t^2$

(a) Find the velocity of the particle when t = 3 seconds   3 marks

(b) Find the acceleration at this point in time   3 marks

(c) A parabola is given by the equation $y = x^2 + 2x - 8$. Write this equation in the form $y = (x + a)^2 + b$ and find the values of 'a' and 'b'.   6 marks

Practice Test 2 Section A   52 Marks

(1) In the triangle below find the length of the side AB. You are given that angle ACB = 32°, AC = 6.2cm and CB = 12.4 cm

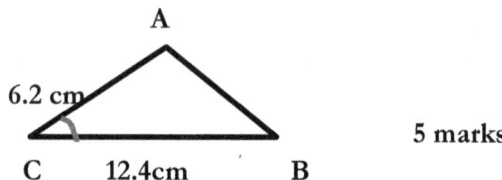

5 marks

(2) (a) If $\tan(x) = \frac{-5}{3}$ for $-90° \leq x \leq 90°$. Find the value(s) of x in this range.   3 marks

(b) Show that $\dfrac{(1+\cos(\theta))(1-\cos(\theta))}{\sin(\theta)} = \sin(\theta)$ 3 marks

(3) A particle moves distance s metres in time t seconds given by the equation $S = \dfrac{3t^3}{2} - t^2$

(a) Find the velocity of the particle when t = 3 seconds

3 marks

(b) Find the acceleration at this point in time   3 marks

96

(4) A curve is given by the equation $y = 2x^3 + 9.5x^2 + 10x + 8$
   (a) Find the stationary points of this curve   4 marks

   (b) Determine if this is a maximum or minimum turning point   3 marks

(5) (a) Find the tangent to the curve $y = 3x^2 - 2x + 1$ at the point $x = 2, y = 9$ 4 marks

   (b) Find the equation of the normal to this curve at the same point   3 marks

(6) A circle has an equation $y = x^2 + y^2 - 6x - 8y = 9$
   (a) Re-write the equation in the form $(x - a)^2 + (y - b)^2 = c$ 5 marks
   (b) Find the radius and the centre of this circle   2 marks

(7) (a) Solve the inequality $2x^2 - x - 1 \geq 0$ 4 marks

   (b) Represent the condition $-2 < x \leq 1$ on a number line

   2 marks

(8) Find the intersection points of the quadratic equation $y = x^2 - x - 12$ with the linear equation $y = 5x + 4$ 3 marks

(9) A curve passes through the point P(1, 0) and has gradient given by $\frac{dy}{dx} = 3x^2 - 2x + 1$. Find the equation of the curve.

   4 marks

(10) If $y = \frac{x-b}{\sqrt{t^2 - m}}$ make t the subject   3 marks

Section B   48 Marks

(11) The graph below shows the two curves $f(x) = x^2 + 5$ and $g(x) = 55 - x^2$

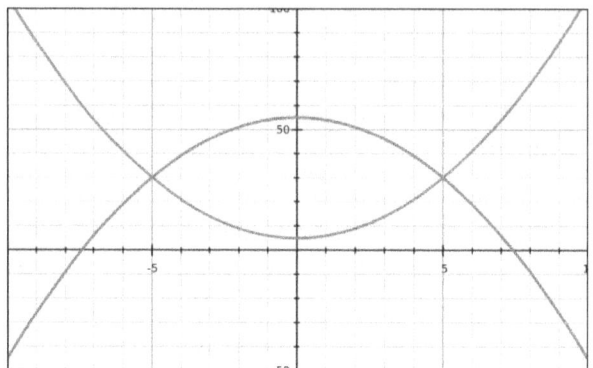

Y – axis

X - axis

(a) Find the co-ordinates of the two intersection points.
2 marks

(b) Find the area enclosed between the two curves
10 marks

(12) (a) A bag consists of 8 red beads and 2 green beads of the same size and shape. A bead is picked at random and then replaced. What is the probability of picking 3 green balls in 50 similar trials? **5 marks**

(b) What is the probability of picking at least 1 red ball in 50 similar trials? **7 marks**

(13) The objective function $z = 2x + 3y$ given is subject to

$x + 3y \leq 21$, $3x - y \geq 0$, and $x - 2y \leq 3$.

(a) Plot the feasible region given by the constraints above in the
graph below 5 marks

(b) Find the intersecting points of the constraints (the corner
points of this region) 5 marks

(c) Maximise the objective function z = 2x + 3y 2 marks

(14) (a) Solve the equation tan(θ) = -1 where $0 \leq \theta \leq 360°$

4 marks

(b) Solve the equation $3(1 - \sin(x)^2) + 2\sin(x) = 2$ for $0 \leq x \leq 360°$
8 marks

Answers to Practice Test 1 Section A and B

1 (a) Answer: y = 2x + 3

Method: At x = 1 the gradient ($\frac{dy}{dx}$) = 4x − 2 = 4×1 -2 = 2

Hence the equation of the tangent at (1, 5) is y − 5 = 2(x - 1)

$\Longrightarrow$ y − 5 = 2x − 2 $\Longrightarrow$ y = 2x + 3

(b) Answer: 2y = 11 − x or y = 5.5 - $\frac{x}{2}$

Method: Gradient of normal can be found using the fact that m1×m2 = -1 (gradient of a line × gradient of its perpendicular = -1)

This means the gradient of the normal is $\frac{-1}{2}$. Hence equation of normal at the same point is y − 5 = $\frac{-1}{2}$ (x − 1)

$\Longrightarrow$ 2y − 10 = -x + 1 $\Longrightarrow$ 2y = 10 − x or y = 5.5 - $\frac{x}{2}$

2 Answer: x ≤ 9

Method: Expanding the bracket to get 3x + 21 ≥ 5x + 3, simplify to get -2x ≥ -18 $\Longrightarrow$ x ≤ 9 (remember when you divide both sides of an inequality with -1, you need to change the sign of the inequality.

3 Answers: x = 41.81° and by symmetry x = (180 − 41.81) = 138.19°

Method: Sin(x) = $\frac{2}{3}$ =0.6666.... **Answers:** x = 41.81° and by symmetry x = (180 − 41.81) = 138.19°

(b) See proof below:

Method: In a right angled triangle below if $\sin(x) = \frac{2}{3} = \frac{opp}{hyp}$

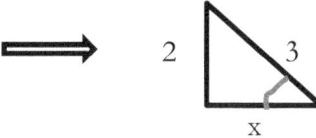

Using Pythagoras theorem we have $9 = 4 + x^2$

⟹ $x = \sqrt{5}$ ⟹ $\tan(x) = \frac{2}{\sqrt{5}}$

4 (a) Answer = (2, 5)

Method: Mid-point of PQ are: $\frac{1+3}{2}, \frac{3+7}{2}$ Hence mid-point is 2,5

(b) Answer: y = 2x + 1

Method: Equation of a straight line given a point it goes through and its gradient is: $y - y_1 = m(x - x_1)$. Using P as a point that it goes through ⟹ $y - 3 = m(x - 1)$

The gradient $m = \frac{7-3}{3-1} = \frac{4}{2} = 2$ ⟹ the equation is

$y - 3 = 2(x - 1)$ ⟹ $y - 3 = 2x - 2$ ⟹ $y = 2x + 1$

(c) Answer: the gradient of the line that is perpendicular to this line is $\frac{-1}{2}$

Method: To find the gradient of the line that is perpendicular to the equation of the line we have just found use the fact that the product of their gradients = -1. That is $m_1 \times m_2 = -1$. Since the gradient of the line we found is 2 ⟹ $2 \times m_2 = -1$. Hence $m_2 = \frac{-1}{2}$

5 (a) Answer: Stationary points occur at $(\frac{1}{3}, 4\frac{2}{3})$

Method:

Stationary point occurs when $\frac{dy}{dx} = 0$

$\Rightarrow 6x - 2 = 0 \Rightarrow x = \frac{1}{3}$ the corresponding value of y is

$3 \times (\frac{1}{3})^2 - 2 \times \frac{1}{3} + 5 = \frac{1}{3} - 2 \times \frac{1}{3} + 5 = -\frac{1}{3} + 5 = 4\frac{2}{3}$

Answer: Stationary points occur at $(\frac{1}{3}, 4\frac{2}{3})$

(b) Answer: Stationary point is a minimum

Method 1: Since the coefficient of x^2 in f(x) =3 which is positive this implies that the stationary point is a minimum.

Method 2: Since $\frac{d^2y}{dx^2} = 6 \Rightarrow \frac{d^2y}{dx^2} > 0$, then this stationary point is a minimum

6 (a) Answer: $y^{\frac{-4}{15}} + y^{\frac{11}{15}}$

Method: Multiply out the bracket to get $y^{\frac{6-10}{15}} + y^{\frac{6+5}{15}}$

(b) Answer:: $\frac{16-x}{x^2 - 9}$

Method: The common denominator is $x^2 - 9$

$\Rightarrow \frac{2}{x-3} - \frac{3}{x+3} + \frac{1}{x^2-9} = \frac{2(x+3)-3(x-3)+1}{x^2-9} = \frac{2x+6-3x+9+1}{x^2-9}$

$= \frac{16-x}{x^2-9}$

7 Answer: $x = \frac{1}{3}$ or $x = \frac{-1}{2}$

Method: Factorise $6x^2 + x - 1 = 0$, to get $(3x - 1)(2x + 1) = 0$

$\Rightarrow 3x = 1$ or $2x = -1 \Rightarrow x = \frac{1}{3}$ or $x = \frac{-1}{2}$

8(a) **Answer: Centre is (3, 4) and diameter 14 units**

Method:
The equation of a circle is given by $(x-a)^2+(y-b)^2=r^2$
Where the centre is (a, b) and radius is r units. This means in the equation given the centre is (3, 4) and radius 7 or diameter 14 units.

(b) Answer is shown by the sketch below

Method: We know the centre is at (3, 4) and the radius is 7 units hence the

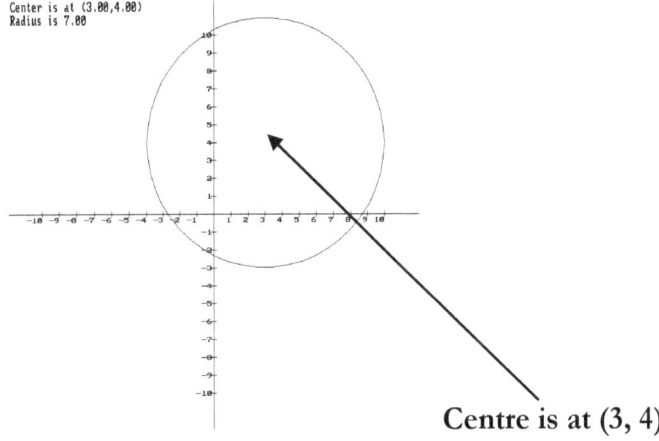

Centre is at (3, 4)

9 **Answer: x = 1+ $\sqrt{2}$ or x = 1 - $\sqrt{2}$**

Method: Solve the simultaneous equations (linear and quadratic)

⟹ $x + 2 = x^2 - x + 1$ ⟹ $x^2 - 2x - 1 = 0$

⟹ Using the quadratic formula: $x = \frac{-b \pm \sqrt{b^2 - 4ac}}{2a}$

⟹ $x = \frac{2 \pm \sqrt{2 \times 2 - 4 \times 1 \times (-1)}}{2 \times 1} = \frac{2 \pm \sqrt{4+4}}{2} = \frac{2 \pm \sqrt{8}}{2} = \frac{2 \pm 2\sqrt{2}}{2}$

$= 1 \pm \sqrt{2}$

10 Answer: See proof below

Method: From (i) $a = \frac{v-u}{t}$ substitute for a in (ii)

⟹ $s = ut + \frac{1}{2}(\frac{v-u}{t})t^2$ ⟹ $s = ut + \frac{1}{2}(v-u)t$

⟹ $s = \frac{2ut + vt - ut}{2}$ ⟹ $s = \frac{t(u+v)}{2}$ as required

11 (a) Answer: 3

Method: Substitute $x = 3$ in $f(x)$ $\implies$ $f(x) = 2\times27 - 9 - 54 + 12 = 3$

(b) Answer: See long division below

Method: Do normal long division

$$\begin{array}{r}
2x^2 + 5x - 3 \text{ remainder } 3 \\
x - 3 \overline{\smash{\big)} 2x^3 - x^2 - 18x + 12} \\
\underline{2x^3 - 6x^2} \\
5x^2 - 18x \\
\underline{5x^2 - 15x} \\
-3x + 12 \\
\underline{-3x + 9} \\
3
\end{array}$$

(c) Answer = -84

Method: work out $g(-2)$ by substituting $x = -2$ in equation $g(x)$

$\implies$ $(-2 -1)(2\times-8 - 4 + 36 + 12) = -3(-16 -4 +48) = -3(28)$. **Answer = -84**

(d) Answer fg(1)) = 12

Method: First work out $g(1)$. This $=0$ $\implies$ $f(0) = 12$ Hence $fg(1) = 12$

12 (a) Answer: 0.00028

Method: Probability of not landing heads = 0.58

$\implies$ Probability of not landing heads for all 15 throws $(0.58)^{15} = 0.00028$

(b) Answer: = 0.99665

Method: Probability that it lands heads twice =

$1 - p(\text{no heads}) - p(\text{one head})$

$= 1 - \binom{15}{0} \times (0.42)^0 \times (0.58)^{15} - \binom{15}{1} \times (0.42)^1 \times (0.58)^{14}$

$= 1 - 0.00028 - 0.00307 = 0.99665$. **Note you already worked out the probability of p(0 heads) in (a)**

13 (a) Answer: (-1.5, -1) and (1, 4)

Method: Solve the two equations above:

$\implies 2x^2 + 3x - 1 = 2x + 2 \quad \implies 2x^2 + x - 3 = 0$

$\implies (2x + 3)(x - 1) = 0 \quad \implies x = -1.5 \text{ or } 1$

Substituting for x in the equation $y = 2x + 2$, we can find the co-ordinates of the two intersecting points.

(b) Answer: y = 2x + 1

Method: A parallel line that goes through (-2, -3) will have the same gradient as $y = 2x + 3$

Hence the equation of this parallel line is $y - (-3) = 2(x - (-2))$

$\implies y + 3 = 2x + 4 \implies y = 2x + 1$

(c) **Answer: Minimum**

Method 1: Since the leading coefficient of $y = 2x^2 + 3x - 1$ is positive then it means the turning point is a minimum.

Method 2

Turning point occurs when $\frac{dy}{dx} = 0$ since $\frac{dy}{dx} = 4x + 3$

If $\frac{d^2y}{dx^2} > 0$, then this turning point is a minimum. Since $\frac{d^2y}{dx^2} = 4$

$\Longrightarrow$ The turning point is a minimum.

14 (a) Answer: 34.5 ms^{-1}

Method: velocity, $v = \frac{ds}{dt} = \frac{9t^2}{2} - 2t = \frac{9 \times 9}{2} - 2 \times 3 = 40.5 - 6 = 34.5$ ms^{-1}

(b) Answer = 25 ms^{-2}

Method: acceleration $= \frac{dv}{dt} = \frac{18t}{2} - 2 = \frac{18 \times 3}{2} - 2 = 27 - 2$

(c) Answer: a = 1 and b = -9

Method: Expand $y = (x + a)^2 + b$ and equate it to

$y = x^2 + 2x - 8 \Longrightarrow y = x^2 + 2ax + a^2 + b = x^2 + 2x - 8$

$\Longrightarrow 2a = 2$ and $a^2 + b = -8 \Longrightarrow a = 1$ and $1 + b = -8 \Longrightarrow b = -9$

Answers to Practice Test 2   Section A and B

(1) Answer: AB = 7.86 cm

Method: in the triangle below use the cosine rule:

$$c^2 = a^2 + b^2 - 2ab\cos C$$

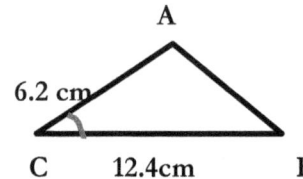

Since $c^2 = a^2 + b^2 - 2ab\cos C$

⟹ $AB^2 = 12.4^2 + 6.2^2 - 2\times 6.2 \times 12.4 \times \cos 32$

⟹ $AB^2 = 153.76 + 38.44 - 130.4 = 61.8$

⟹ AB = 7.86 cm

2 (a) Answer: x = -59.04°

Method: $Tan^{-1}(-5/3) = -59.04°$

(b) See method below

Method: cross multiply by $\sin(\theta)$ to get :

$(1 + \cos(\theta))(1 - \cos(\theta)) = \sin^2(\theta)$

⟹ $1 - \cos^2(\theta) = \sin^2(\theta)$ ⟹ $1 = \sin^2(\theta) + \cos^2(\theta)$

Since $\sin^2(\theta) + \cos^2(\theta) = 1$ ⟹ $\dfrac{(1+\cos(\theta))(1-\cos(\theta))}{\sin(\theta)} = \sin(\theta)$

3 (a) Answer: 34.5ms^{-1}

Method: velocity (v) = $\dfrac{ds}{dt}$ = = $\dfrac{9t^2}{2} - 2t$

⇨ When t = 3 seconds v = $\dfrac{9 \times 9}{2}$ - 2×3 = 40.5 -6 = 34.5ms^{-1}

(b) Answer: 25ms^{-2}

Method: acceleration (a) = $\dfrac{dv}{dt}$ = $\dfrac{18t}{2} - 2$ =

⇨ When t =3 seconds, = $\dfrac{dv}{dt}$ = $\dfrac{18 \times 3}{2} - 2$ = 25ms^{-2}

4(a) Answer: x = -2.5, y = 11.14

Method: Stationary points occur when $\dfrac{dy}{dx} = 0$

⇨ $\dfrac{dy}{dx} = 6x^2 + 19x + 10 = 0$

⇨ (3x + 2)(2x + 5) =0 ⇨ x = $\dfrac{-2}{3}$ or x = - 2.5

⇨ Corresponding values of y can be found by substituting the values of x in the equation y = $2x^3 + 9.5x^2 + 10x + 8$

⇨ When x = $\dfrac{-2}{3}$, y = 4.96 and when x = -2.5, y = 11.14

(b) **Answer:** when $x = \frac{-2}{3}$, turning point is a minimum and when $x = 2.5$ it is a maximum

Method: If $\frac{d^2y}{dx^2} > 0$, then this turning point is a minimum similarly if $\frac{d^2y}{dx^2} < 0$ then the turning point is a maximum.

$\frac{d^2y}{dx^2} = 12x + 19 \implies$ when $x = \frac{-2}{3}$, $\frac{d^2y}{dx^2} = 12 \times \frac{-2}{3} + 19 = 11$

$\implies \frac{d^2y}{dx^2} > 0$ hence at this point the turning point is a minimum

Also when $x = -2.5$ then, $\frac{d^2y}{dx^2} = 12 \times -2.5 + 19 = -30 + 19 = -11$

$\implies$ When $x = 2.5$ the turning point is a maximum

5 (a) **Answer:** $y = 10x - 11$

Method: First find the gradient of $y = 3x^2 - 2x + 1$

$\implies \frac{dy}{dx} = 6x - 2 \implies$ at $x = 2$, $\frac{dy}{dx} = 12 - 2 = 10$

$\implies$ The equation of the tangent at (2, 9) can be found using the formula $y - y1 = m(x - x1)$

$\implies y - 9 = 10(x - 2) \implies y - 9 = 10x - 20 \implies y = 10x - 11$

Hence the equation of the tangent at this point is **y = 10x -11**

(b) **Answer:** $10y = 92 - x$

Method: To find the equation of the normal first find the gradient of the normal (line that is perpendicular to the tangent)

Using the fact that **m1×m2 = -1** $\implies$ 10×m2 = -1

$\Rightarrow m_2 = \frac{-1}{10}$ $\Rightarrow y - 9 = \frac{-1}{10}(x - 2)$

$\Rightarrow 10y - 90 = -x + 2$ $\Rightarrow 10y = 92 - x$

Hence the equation of the normal at this point is $10y = 92 - x$

6(a) Answer: $(x - 3)^2 + (y - 4)^2 = 16$

Method: Equate $y = x^2 + y^2 - 6x - 8y = 9$ to $(x - a)^2 + (y - b)^2 = c$. Now expanding the brackets we get $x^2 - 2ax + a^2 + y^2 - 2by + b^2 = c$

$\Rightarrow -6 = -2a$ and $-8 = $ $\Rightarrow$ $a = 3$ and $b = 4$

Finally since $a^2 + b^2 - c = 9$ $\Rightarrow$ $9 + 16 - c = 9$

$\Rightarrow c = 16$ $\Rightarrow$ the equation required is $(x - 3)^2 + (y - 4)^2 = 16$

(b) Answer: Centre is at the point (3, 4) with radius is 4 units

Method: In the equation of a circle below:

$(x - a)^2 + (y - b)^2 = r^2$, a, b are the co-ordinates of the centre and r is the radius. Now compare this to $(x - 3)^2 + (y - 4)^2 = 16$

$\Rightarrow$ The centre is at the point (3, 4) and the radius is 4 units

7 (a) Answer: $x < \frac{-1}{2}$ or $x \geq 1$

Method: Factorise $2x^2 - x - 1$ to get $(2x + 1)(x - 1)$

$\Rightarrow (2x + 1)(x - 1) = 0$ $\Rightarrow$ either $2x + 1 = 0$ or $x - 1 = 0$

$\Rightarrow$ $x = \frac{-1}{2}$ or $x = 1$ **(this gives us the intercepts)**

⇒ If $2x^2 - x - 1 \geq 0$ then $x \leq \frac{-1}{2}$ or $x \geq 1$

(b) **Answer: As per line shown below:**

Method: Draw the line with circles to show the inequality required.

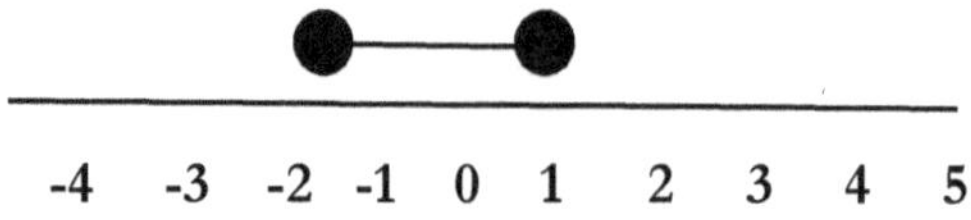

```
            ●———●
_____
   -4  -3  -2  -1  0  1  2  3  4  5
```

8 Answer: x = 8 or x = -2

Method: Solve the simultaneous equations (linear and quadratic)

$y = x^2 - x - 12$ and $y = 5x + 4$

⇒ $x^2 - x - 12 = 5x + 4$

⇒ $x^2 - 6x - 16 = 0$

⇒ $(x - 8)(x + 2) = 0$

⇒ $x = 8$ or $x = -2$

9 Answer: $y = x^3 - x^2 + x - 1$

Method: Since we know $\frac{dy}{dx}$, we need to integrate to find y. That is we need to find $\int (3x^2 - 2x + 1) dx$

⇒ $y = \frac{3x^3}{3} - \frac{2x^2}{2} + x + c$ ⇒ $y = x^3 - x^2 + x + c$

Since this equation goes through P(1,0) we can find the value of c by substituting the values of P in the equation $y = x^3 - x^2 + x + c$

⇨ $0 = 1 - 1 + 1 + c$ ⇨ $c = -1$

Hence the equation of the curve is $y = x^3 - x^2 + x - 1$

10 Answer: $t = \sqrt{\dfrac{(x-b)^2}{y^2} + m}$

Method: Square both sides to get $y^2 = \dfrac{(x-b)^2}{t^2 - m}$

Cross multiply by $t^2 - m$ to get $y^2(t^2 - m) = (x - b)^2$

Divide both sides by y^2 ⇨ $t^2 - m = \dfrac{(x-b)^2}{y^2}$

Add m to both sides ⇨ $t^2 = \dfrac{(x-b)^2}{y^2} + m$

⇨ $t = \sqrt{\dfrac{(x-b)^2}{y^2} + m}$

(N.B. the value of t can strictly be + or – the answer given)

(11) (a) Answer: (-5, 30) and (5, 30)

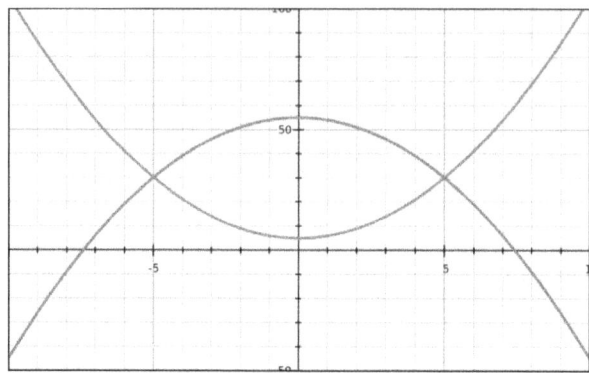

Method: Solve the simultaneous quadratic equations f(x) and g(x)

$\implies x^2 + 5 = 55 - x^2 \implies 2x^2 = 50 \implies x = +5$ or -5

By substituting the values of x in f(x) we can find the y co=ordinates

$\implies$ The two intersecting points are (-5, 30) and (5, 30)

(b) Answer: Area = $333\frac{1}{3}$ sq. units

Method: Area required is area of the inverted U-shaped curve minus the area of the U-shaped curve between the intervals shown x = -5 and x = 5

Area under inverted U-shaped curve between x =-5 and x =5 is

$\int_{-5}^{5}(55 - x^2) \, dx = \left[55x + \frac{x^3}{3}\right]_{-5}^{5} = (275 - \frac{125}{3}) - (-275 - (-\frac{125}{3}))$

$= 550 - \frac{250}{3} = 550 - 83\frac{1}{3} = 466\frac{2}{3}$ square units

Similarly the area under the U-shaped curve between these intervals is: $\int_{-5}^{5}(x^2 + 5) \, dx = \left[\frac{x^3}{3} + 5x\right]_{-5}^{5} = (\frac{125}{3} + 25) - (\frac{-125}{3} - 25) = \frac{250}{3} + 50 = 83\frac{1}{3} + 50 = 133\frac{1}{3}$ square units

So area between the two curves is $= 466\frac{2}{3} - 133\frac{1}{3} = 333\frac{1}{3}$ sq. units

(12) (a) Answer: 0.004371

Method: Using the binomial Probability distribution, the probability of picking 3 green beads in 50 similar trials =

$\binom{50}{3} \times (0.2)^3 \times (0.8)^{50-3} = \frac{50 \times 49 \times 48 \times 47!}{47! \times 3!} \times 2.23007\text{E-}07 = 0.004371$

(b) Answer: nearly 1.

Method: Probability of picking at least one red ball = 1 – prob(0 red balls)

$= 1 - \binom{50}{0} \times (0.8)^0 \times (0.2)^{50} = 1 - (0.2)^{50} = $ nearly 1. Almost certain to pick at least one red ball.

(13) (a) Answer: The feasible region is shown below:

Method: Plot each linear equation given as constraints. The feasible region is the triangular region shown by the arrow.

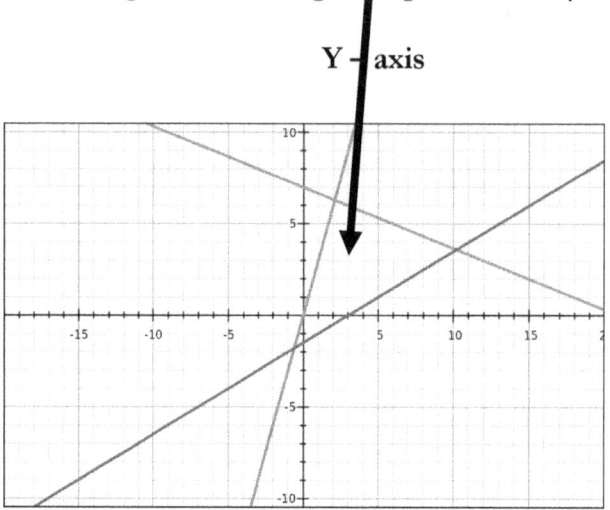

(b) Answer: The vertices are (2.1, 6.3), (10.2, 3.6) and (-0.6, -1.8).

Method solve the appropriate equations (e.g. solve $x + 3y \leq 21$ and $3x - y \geq 0$ to find $x = 2.1$, $y = 6.3$, similarly solve the other simultaneous pairs to find the intersecting points or the vertices of the feasible region.) The vertices are **(2.1, 6.3), (10.2, 3.6) and (-0.6, -1.8).** (Note that the optimum value of the objective function will occur at one of the vertices of the feasible region)

(c) **Answer: 31.2**

Method: The optimum (in this case the maximum) value of the objective function occurs at **((10.2, 3.6)** ⟹ the maximum value of the objective function is $2 \times 10.2 + 3 \times 3.6 = 20.4 + 10.8 = 31.2$

(14) (a) Answer: θ = 135° and 315°

Method: from the graph of $y = \tan(\theta)$ shown below we can see that there are two values between the range given. Also we can work out $\theta = tan^{-1}(-1)$ exactly using the appropriate trig function. Since if $\tan(\theta) = 1$ then $\theta = 45°$, by symmetry of the graph the values of θ when $\tan(\theta) = -1$ are $(180 - 45) = \mathbf{135°}$ and $(360 - 45) = \mathbf{315°}$

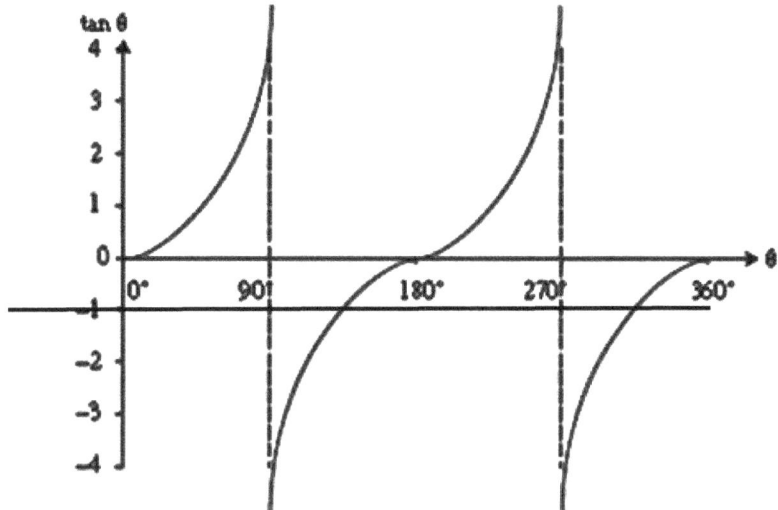

(b) Answer: x = 90°, 199.47° and 340.53°

(To solve the equation $3(1 - \sin(x)^2) + 2\sin(x) = 2$ for $0 \leq x \leq 360°$ see method below: If you find it easier let $\sin(x) = y$ and then solve the resulting quadratic equation in y.)

Method: To solve the equation $3(1 - \sin(x)^2) + 2\sin(x) = 2$ for $0 \leq x \leq 360°$, Expanding the bracket we have $3 - 3\sin(x)^2 + 2\sin(x) = 2$.

⇒ $3\sin(x)^2 - 2\sin(x) - 1 = 0$ ⇒ $3\sin(x) + 1)(\sin(x) - 1) = 0$

⇒ $\sin(x) = -0.33333...$ or $\sin(x) = 1$

From the sine curve (and using trig functions in a calculator as appropriate) we can see that $\sin(x) = 0.333333$, when $x = (180 +$

19.47) = **199.47 or** (360 − 19.47) = **340.53°** and sin(x) = **1** when x =90°. Hence the solutions are **90°, 199.47° and 340.53°**

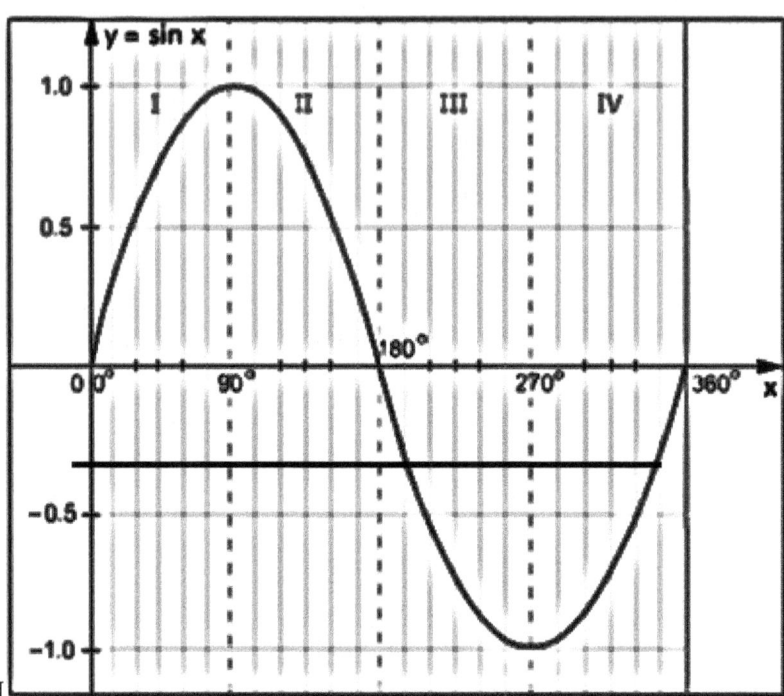

www.ingramcontent.com/pod-product-compliance
Lightning Source LLC
Chambersburg PA
CBHW051548170526
45165CB00002B/926